William Shakespeare, Bessie Mayou

Natural History of Shakespeare

Being selection of flowers, fruits and animals

William Shakespeare, Bessie Mayou

Natural History of Shakespeare
Being selection of flowers, fruits and animals

ISBN/EAN: 9783337238537

Printed in Europe, USA, Canada, Australia, Japan

Cover: Foto ©berggeist007 / pixelio.de

More available books at **www.hansebooks.com**

NATURAL HISTORY

OF

SHAKESPEARE

ALL the Quotations in this book are taken from the text of Johnson and Stevens, and the "Handy Volume" edition of Shakespeare.

B. M.

1877.

OF

SHAKESPEARE

BEING SELECTIONS OF

FLOWERS, FRUITS, AND ANIMALS

ARRANGED BY

BESSIE MAYOU

One touch of Nature makes the whole world kin.
TROILUS AND CRESSIDA, Act iii. Scene 3.

EDWIN SLATER

(BOOKSELLER AND PUBLISHER TO H.R.H. THE PRINCE OF WALES)

16 ST. ANN'S SQUARE, MANCHESTER

[*All rights reserved.*]

DEDICATED BY PERMISSION

TO

THE RIGHT HONOURABLE

THE BARONESS BURDETT-COUTTS.

PREFACE.

In publishing the present work, I feel that some short explanation is necessary. I was first led to take up the Natural History of Shakespeare from a short paragraph I saw in the "Garden" a few years ago, and was struck with the very large number of flowers, fruits, vegetables, etc., mentioned; as we must remember Shakespeare's object was not to write of trees and plants, but to use them as illustrations, and also that, three centuries ago, very little was known of botany, more especially of English wild flowers. With regard to the Animal Kingdom, it is merely a continuation of the former, as I had no idea of ever completing my work when I commenced it two years ago.

In conclusion, let me say that love, not presumption, prompted me throughout, and let Shakespeare speak for me—

> "He that of greatest works is Finisher,
> Oft does them by the weakest minister."

BESSIE MAYOU.

Cheetham Hill.

GARDEN FLOWERS.

Here's a few flowers.
CYMBELINE, Act iv. Scene 2.

ROSE.

Oberon. Quite over-canopied with lush woodbine,
With sweet musk roses, and with eglantine:

Titania. Come, sit thee down upon this flowery bed,
While I thy amiable cheeks do coy,
And stick musk roses in thy sleek smooth head,
And kiss thy fair large ears, my gentle joy.
A MIDSUMMER NIGHT'S DREAM, Act ii. Scene 1;
Act iv. Scene 1.

Don John. I had rather be a canker in a hedge than a rose in his grace;
MUCH ADO ABOUT NOTHING, Act i. Scene 3.

Duke. . . . let thy love be younger than thyself,
Or thy affection cannot hold the bent:
For women are as roses; whose fair flower,
Being once display'd, doth fall that very hour.
TWELFTH NIGHT, Act ii. Scene 4.

Biron. At Christmas I no more desire a rose,
Than wish a snow in May's new-fangled shows;

Princess. Will they return?
Boyet. They will, they will,
And leap for joy, though they are lame with blows:
Therefore, change favours; and, when they repair,
Blow like sweet roses in this summer air.
Princess. How blow? how blow? speak to be understood.
Boyet. Fair ladies, mask'd, are roses in their bud:
Dismask'd their damask sweet commixture shown,
Are angels vailing clouds, or roses blown.

Love's Labour's Lost, Act i. Scene 1;
Act v. Scene 2.

Touchstone. He that sweetest rose will find,
Must find love's prick and Rosalind.

As You Like It, Act iii. Scene 2.

Petrucio.   I'll say, she looks as clear
As morning roses newly wash'd with dew;

Taming of the Shrew, Act ii. Scene 1.

Constance. . . . at thy birth, dear boy,
Nature and Fortune join'd to make thee great:
Of Nature's gifts thou may'st with lilies boast,
And with the half-blown rose.

King John, Act iii. Scene 1.

Queen. But soft, but see, or rather do not see,
My fair rose wither; yet look up; behold;
That you in pity may dissolve to dew,
And wash him fresh again with true-love tears.

King Richard II., Act v. Scene 1.

Hostess. Your colour, I warrant you, is as red as any rose.

KING HENRY IV., Part II. Act ii. Scene 4.

Plantagenet. Let him that is a true-born gentleman,
And stands upon the honour of his birth,
If he suppose that I have pleaded truth,
From off this brier pluck a white rose with me.

Somerset. Let him that is no coward, nor no flatterer
But dare maintain the party of the truth,
Pluck a red rose from off this thorn with me.

Warwick. I love no colours; and, without all colour
Of base insinuating flattery,
I pluck this white rose with Plantagenet.

Suffolk. I pluck this red rose with young Somerset;
And say withal, I think he held the right.

Vernon. Stay, lords and gentlemen; and pluck no more,
Till you conclude—that he upon whose side
The fewest roses are cropp'd from the tree,
Shall yield the other in the right opinion.

Somerset. Good master Vernon, it is well objected:
If I have fewest I subscribe in silence.

Plantagenet. And I.

Vernon. Then, for the truth and plainness of the case,
I pluck this pale and maiden blossom here,
Giving my verdict on the white rose side.

Somerset. Prick not your finger as you pluck it off,
Lest, bleeding, you do paint the white rose red,
And fall on my side so, against your will.

Vernon. If I, my lord, for my opinion bleed,
Opinion shall be surgeon to my hurt,
And keep me on the side where still I am.

Somerset. Well, well, come on; who else?

Lawyer (to Somerset). Unless my study and my books be false,
The argument you held was wrong in you;
In sign whereof I pluck a white rose too.

Plantagenet. Now, Somerset, where is your argument?

Somerset. Here, in my scabbard; meditating that
Shall dye your white rose in a bloody red.

Plantagenet. Meantime, your cheeks do counterfeit our roses;
For pale they look with fear, as witnessing
The truth on our side.

Somerset. No, Plantagenet,
'Tis not for fear, but anger,—that thy cheeks
Blush for pure shame, to counterfeit our roses;
And yet thy tongue will not confess thy error.

Plantagenet. Hath not thy rose a canker, Somerset?

Somerset. Hath not thy rose a thorn, Plantagenet?

Plantagenet. Ay, sharp and piercing, to maintain his truth;
Whiles thy consuming canker eats his falsehood.

Somerset. Well, I'll find friends to wear my bleeding roses,
That shall maintain what I have said is true,
Where false Plantagenet dare not be seen.

Plantagenet. Now, by this maiden blossom in my hand,
I scorn thee and thy faction, peevish boy.

. . . .

Warwick. Meantime, in signal of my love to thee,
. . . . Will I upon thy party wear this rose;
And here I prophesy,—This brawl to-day,
Grown to this faction in the Temple garden,

Shall send, between the red rose and the white,
A thousand souls to death and deadly night.
>> KING HENRY VI., Part I. Act ii. Scene 4.

York. Then will I raise aloft the milk-white rose,
With whose sweet smell the air shall be perfumed
>> KING HENRY VI., Part II. Act i. Scene 1.

King Henry. The red rose and the white are on his face,
The fatal colours of our striving houses :
>> KING HENRY VI., Part III. Act ii. Scene 5.

Richmond. And then, as we have ta'en the sacrament,
We will unite the white rose and the red ;
>> KING RICHARD III., Act v. Scene 4.

Antony. . . . tell him, he wears the rose
Of youth upon him;
>> ANTONY AND CLEOPATRA, Act iii. Scene 11.

Othello. . . . When I have pluck'd thy rose,
I cannot give it vital growth again,
It needs must wither ;——
>> OTHELLO, Act v. Scene 2.

Juliet. What's in a name? that which we call a rose,
By any other name would smell as sweet ;
>> ROMEO AND JULIET, Act ii. Scene 2.

There will I make thee a bed of roses,
With a thousand fragrant posies,
>> PASSIONATE PILGRIM, xviii.

Roses have thorns, and silver fountains mud;
> Sonnet XXXV.

The rose looks fair, but fairer we it deem
For that sweet odour which doth in it live.
The canker-blooms have full as deep a dye
As the perfumèd tincture of the roses,
Hang on such thorns, and play as wantonly
When summer's breath their masked buds
    discloses;
But, for their virtue only is their show,
They live unwoo'd, and unrespected fade;
Die to themselves. Sweet roses do not so;
Of their sweet deaths are sweetest odours made:
> Sonnet LIV.

LILY.

Princess. . . . by my maiden honour, yet as pure
As the unsullied lily.
> Love's Labour's Lost, Act v. Scene 2.

Constance. Of Nature's gifts thou mayst with lilies boast,

Salisbury. To gild refinèd gold, to paint the lily.
. . . . . . Is wasteful, and ridiculous excess.
> King John, Act iii. Scene 1; Act iv. Scene 2.

Queen Katherine.   like the lily,
That once was mistress of the field and flourish'd,
I'll hang my head and perish.

Cranmer. A most unspotted lily shall she pass
To the ground, and all the world shall mourn her.
   KING HENRY VIII., Act iii. Scene 1 ; Act v. Scene 5.

Perdita. . . . lilies of all kinds,
The flower-de-luce being one !
   WINTER'S TALE, Act iv. Scene 3.

Troilus. . . . give me swift transportance to those fields
Where I may wallow in the lily beds
   TROILUS AND CRESSIDA, Act iii. Scene 2.

Guiderius.   O sweetest, fairest lily !
My brother wears thee not the one-half so well,
As when thou grew'st thyself.
   CYMBELINE, Act iv. Scene 2.

Titus. . . . fresh tears
Stood on her cheeks, as doth the honey-dew
Upon a gather'd lily almost wither'd.
   TITUS ANDRONICUS, Act iii. Scene 1.

Lilies that fester smell far worse than weeds.
   SONNET XCIV.

Nor did I wonder at the lily's white,
   SONNET XCVIII.

The lily I condemned for thy hand,
   SONNET XCIX.

CARNATION.

Perdita. . . . the year growing ancient,—
Not yet on summer's death, nor on the birth
Of trembling winter,—the fairest flowers o' the season
Are our carnations,
   A WINTER'S TALE, Act iv. Scene 3.

MARIGOLD.

Perdita. The marigold, that goes to bed with the sun,
And with him rises weeping ;
A Winter's Tale, Act iv. Scene 3.

Marina. . . . I will rob Tellus of her weed,
To strew thy green with flowers ; the yellows, blues,
The purple violets, and marigolds,
Shall as a chaplet hang upon thy grave,
While summer days do last.
Pericles, Act iv. Scene 1.

Her eyes, like marigolds, had sheathed their light,
And canopied in darkness sweetly lay,
Till they might open to adorn the day.
The Rape of Lucrece.

Great princes' favourites their fair leaves spread
But as the marigold at the sun's eye ;
Sonnet XXV.

COLUMBINE.

Longueville. That columbine.
Love's Labour's Lost, Act v. Scene 2.

Ophelia. There's fennel for you, and columbines :—
Hamlet, Act iv. Scene 5.

PANSY.

Ophelia. . . . there is pansies,
that's for thoughts.
Hamlet, Act iv. Scene 5.

LAVENDER.

Perdita. . . . Here's flowers for you;
Hot lavender,
          A WINTER'S TALE, Act iv. Scene 3.

PINK.

Mercutio. . . . I am the very pink of courtesy.
Romeo. Pink for flower.
Mercutio. Right.
          ROMEO AND JULIET, Act ii. Scene 4.

GILLYVORS (WALLFLOWER).

Perdita. . . . the fairest flowers o' the season,
Are our carnations, and streak'd gillyvors,
Which some call nature's bastards: of that kind
Our rustic garden's barren; and I care not
To get slips of them.
   .   .   .   .
Polixenes. Then make your garden rich in gillyvors,
And do not call them bastards.
          A WINTER'S TALE, Act iv. Scene 3.

HOLY THISTLE.

Beatrice. By my troth, I am sick.
Margaret. Get you some of this distilled Carduus Benedictus, and lay it to your heart; it is the only thing for a qualm.

Hero. There thou prick'st her with a thistle.

Beatrice. *Benedictus!* why *Benedictus?* you have some moral in this Benedictus.

Margaret. Moral! no, by my troth, I have no moral meaning; I meant, plain holy-thistle.

Much Ado about Nothing, Act iii. Scene 4.

WILD FLOWERS.

In emerald tufts, flowers purple, blue and white
Like sapphire, pearl, and rich embroidery,
MERRY WIVES OF WINDSOR, Act v. Scene 5.

VIOLET.

Oberon. I know a bank where the wild thyme blows,
Where ox-lips and the nodding violet grows;
MIDSUMMER NIGHT'S DREAM, Act ii. Scene 1.

Duke. That strain again!—it had a dying fall:
Oh, it came o'er my ear like the sweet south
That breathes upon a bank of violets,
Stealing and giving odour!—
TWELFTH NIGHT, Act i. Scene 1.

Perdita. . . . violets, dim,
But sweeter than the lids of Juno's eyes,
Or Cytherea's breath;
A WINTER'S TALE, Act iv. Scene 3.

Duchess. Welcome, my son: who are the violets now
That strew the green lap of the new-come spring?
KING RICHARD II., Act v. Scene 2.

King Henry. . . . I think the king is but a man, as I am; the violet smells to him as it doth to me;
KING HENRY V., Act iv. Scene 1.

Salisbury. To throw a perfume on the violet,
. . . .
Is wasteful, and ridiculous excess.
KING JOHN, Act iv. Scene 2.

Laertes. A violet in the youth of primy nature,
Forward, not permanent, sweet, not lasting,

Ophelia. I would give you some violets; but they withered all, when my father died:—

Laertes. Lay her i' the earth;
And from her fair and unpolluted flesh
May violets spring!
HAMLET, Act i. Scene 3; Act iv. Scene 5; Act v. Scene 1.

Belarius. They are as gentle
As zephyrs, blowing below the violet,
Not wagging his sweet head:
CYMBELINE, Act iv. Scene 2.

When I behold the violet past prime,
SONNET XII.

The forward violet thus did I chide:—
Sweet thief, whence didst thou steal thy sweet that
    smells,
If not from my love's breath?
SONNET XCIX.

DAFFODIL.

When daffodils begin to peer,
With heigh! the doxy over the dale,
Why then comes in the sweet o' the year;
For the red blood reigns in the winter's pale.

Perdita. . . . daffodils,
That come before the swallow dares, and take
The winds of March with beauty;
      A WINTER'S TALE, Act iv. Scene 2 and Scene 3.

PRIMROSE.

Perdita. . . . pale primroses,
That die unmarried, ere they can behold
Bright Phœbus in his strength,
      A WINTER'S TALE, Act iv. Scene 3.

Hermia. And in the wood, where often you and I
Upon faint primrose beds were wont to lie,
      A MIDSUMMER NIGHT'S DREAM, Act i. Scene 1.

Arviragus. . . . thou shalt not lack
The flower that's like thy face, pale primrose;
      CYMBELINE, Act iv. Scene 2.

Q. Margaret. Look pale as primrose
      KING HENRY VI., Part II. Act iii. Scene 2.

Porter. I had thought to have let in some of all professions, that go the primrose way to the everlasting bonfire.—
      MACBETH, Act ii. Scene 2.

OX-LIP.

Oberon. Where ox-lips and the nodding violet grows;
        MIDSUMMER NIGHT'S DREAM, Act ii. Scene 1.

Perdita. . . . bold ox-lips, and
The crown imperial;
        A WINTER'S TALE, Act iv. Scene 3.

DAISY.

Ophelia. There's a daisy :—

Queen. Of crow-flowers, nettles, daisies, and long-purples,
        HAMLET, Act iv. Scene 5 and Scene 7.

When daisies pied, and violets blue,
. . . .
Do paint the meadows
        LOVE'S LABOUR'S LOST, Act v. Scene 2.

Lucius. . . . let us
Find out the prettiest daisied plot we can,
And make him with our pikes and partisans
A grave.
        CYMBELINE, Act iv. Scene 2.

Without the bed her other fair hand was,
On the green coverlet; whose perfect white
Show'd like an April daisy on the grass,
        THE RAPE OF LUCRECE.

LADY-SMOCKS.

Lady-smocks all silver white,
LOVE'S LABOUR'S LOST, Act v. Scene 2.

WOODBINE AND EGLANTINE.

Oberon. Quite over-canopied with lush woodbine,
With sweet musk-roses, and with eglantine :
   Titania. Sleep thou, and I will wind thee in my arms.

. . . .

So doth the woodbine the sweet honeysuckle
Gently entwist ;
MIDSUMMER NIGHT'S DREAM, Act ii. Scene 1 ;
Act iv. Scene 1.

Ursula. The pleasantest angling is to see the fish
Cut with her golden oars the silver stream,
And greedily devour the treacherous bait :
So angle we for Beatrice ; who even now
Is couched in the woodbine coverture :
MUCH ADO ABOUT NOTHING, Act iii. Scene 1.

Arviragus. The leaf of eglantine, whom not to slander,
Out-sweeten'd not thy breath :
CYMBELINE, Act iv. Scene 2.

HONEYSUCKLE.

Hero. . . . bid her steal into the pleached bower,

Where honeysuckles, ripen'd by the sun.
Forbid the sun to enter;—
>> MUCH ADO ABOUT NOTHING, Act iii. Scene 1.

Hostess. O thou honeysuckle villain!
>> KING HENRY IV., Part II. Act ii. Scene 1.

COWSLIP.

Ariel. In a cowslip's bell I lie:
There I couch when owls do cry.
>> TEMPEST, Act v. Scene 1.

Fairy. . . . I serve the fairy queen,
To dew her orbs upon the green:
The cowslips tall her pensioners be;
In their gold coats spots you see;
Those be rubies, fairy favours,
In those freckles live their savours:
I must go seek some dew-drops here,
And hang a pearl in every cowslip's ear.

Thisbe. These yellow cowslip cheeks,
>> MIDSUMMER NIGHT'S DREAM, Act ii.
>> Scene 1; Act. v. Scene 2.

Burgundy. The even mead, that erst brought sweetly forth
The freckled cowslip,
>> KING HENRY V., Act v. Scene 2.

Queen. The violets, cowslips, and the primroses.
Bear to my closet.—

Jachimo. On her left breast
A mole cinque-spotted, like the crimson drops
I' the bottom of a cowslip.
                CYMBELINE, Act i. Scene 5; Act ii. Scene 2.

HARE-BELL.

Arviragus. The azured hare-bell, like thy veins;
                CYMBELINE, Act iv. Scene 2.

LOVE-IN-IDLENESS.

Oberon. Yet mark'd I where the bolt of Cupid fell;
It fell upon a little western flower,—
Before, milk-white, now purple with love's wound,
And maidens call it love-in-idleness.
                MIDSUMMER NIGHT'S DREAM, Act ii. Scene 1.

FERN.

Gadshill. . . . we have the receipt of fern-seed, we walk invisible.

Chamberlain. Nay, I think rather you are more beholden to the night than to fern-seed, for your walking invisible.
                KING HENRY IV., Part I. Act ii. Scene 1.

WEEDS.

Now 'tis the spring, and weeds are shallow-rooted;
Suffer them now, and they'll o'ergrow the garden.
          KING HENRY VI., Part II. Act iii. Scene 1.

NETTLE.

Gonzalo. Had I plantation of this isle, my lord,—
Antonio. He'd sow't with nettle seed.
          TEMPEST, Act ii. Scene 1.

Sir Toby. How now, my nettle of India?
          TWELFTH NIGHT, Act ii. Scene 5.

King Richard. Yield stinging nettles to mine enemies:
          KING RICHARD II., Act iii. Scene 2.

Hotspur. . . . out of this nettle, danger, we pluck this flower, safety.
          KING HENRY IV., Part I. Act ii. Scene 3.

Pandarus. . . . he will weep you, an 'twere a man born in April.
Cressida. And I'll spring up in his tears, an 'twere a nettle against May.
          TROILUS AND CRESSIDA, Act i. Scene 2.

Iago. Our bodies are our gardens; to the which our wills are gardeners: so that if we will plant nettles, . . . why, the power and corrigible authority of this lies in our wills.

OTHELLO, Act i. Scene 3.

Menenius. We call a nettle but a nettle;
And the faults of fools but folly.

CORIOLANUS, Act ii. Scene 1.

COCKLE.

Biron. Allons! Allons!—Sow'd cockle reap'd no corn;

LOVE'S LABOUR'S LOST, Act iv. Scene 3.

Coriolanus. I say again,
In soothing them we nourish 'gainst our senate
The cockle of rebellion,

CORIOLANUS, Act iii. Scene 1.

DARNEL.

Burgundy. . . . her fallow leas
The darnel, hemlock, and rank fumitory,
Doth root upon;

KING HENRY V., Act v. Scene 2.

La Pucelle. Good morrow, gallants! want ye corn for bread?
I think the Duke of Burgundy will fast,
Before he'll buy again at such a rate:
'Twas full of darnel:

KING HENRY VI., Part I. Act iii. Scene 2.

Cordelia. Darnel, and all the idle weeds that grow
In our sustaining corn.——
KING LEAR, Act iv. Scene 4.

FUMITORY.

Cordelia. Crown'd with rank fumiter, and furrow
    weeds,
KING LEAR, Act iv. Scene 4.

Burgundy. . . rank fumitory,
KING HENRY V., Act v. Scene 2.

HARLOCK. HEMLOCK.

Cordelia. . . . harlocks, hemlock, nettles, cuckoo
    flowers,
KING LEAR. Act iv. Scene 4.

3d Witch. Root of hemlock. digg'd i' the dark ;
MACBETH, Act iv. Scene 1.

DOCK.

Gonzalo. Had I plantation of this isle, my lord,—
Antonio. He'd sow't with nettle seed.
Sebastian.   Or docks. or mallows.
THE TEMPEST, Act ii. Scene 1.

Burgundy. . . . and nothing teems
    But hateful docks.
KING HENRY V., Act v. Scene 2.

THISTLE.

Bottom. Monsieur Cobweb ; good monsieur, get

your weapons in your hand, and kill me a red-hipped humble-bee on the top of a thistle ;
MIDSUMMER NIGHT'S DREAM, Act iv. Scene 1.

Burgundy. . . . rough thistles, kecksies, burs, Losing both beauty and utility :
KING HENRY V., Act v. Scene 2.

RUSHES.

Dromio, S. A rush, a hair, a drop of blood, a pin,
COMEDY OF ERRORS, Act iv. Scene 3.

Rosalind. There is none of my uncle's marks upon you : he taught me how to know a man in love ; in which cage of rushes, I am sure, you are not prisoner.
 . . . lean upon a rush,
The cicatrice and capable impressure
Thy palm some moment keeps :
AS YOU LIKE IT, Act iii. Scenes 2 and 5.

Grumio. Where's the cook ? is supper ready, the house trimmed, rushes strewed, cobwebs swept ;

Katharine. And be it moon, or sun, or what you please :
And if you please to call it a rush candle,
Henceforth I vow it shall be so for me.
TAMING OF THE SHREW, Act iv. Scenes 1 and 5.

Bastard. . . . a rush will be a beam
To hang thee on ;
KING JOHN, Act iv. Scene 3.

1st Groom. More rushes, more rushes.
KING HENRY IV., Part II. Act v. Scene 5.

Eros. He's walking in the garden—thus: and spurns
The rush that lies before him;
ANTONY AND CLEOPATRA, Act iii. Scene 5.

Marcius. He that depends
Upon your favours, swims with fins of lead,
And hews down oaks with rushes.

1st Senator. . . . our gates,
Which yet seem shut, we have but pinn'd with rushes;
CORIOLANUS, Act i. Scenes 1 and 4.

Romeo. . . . let wantons, light of heart,
Tickle the senseless rushes with their heels;
ROMEO AND JULIET, Act i. Scene 4.

BURS.

Lucio. . . . I am a kind of bur,
I shall stick.
MEASURE FOR MEASURE, Act iv. Scene 3.

Lysander. Hang off, thou cat, thou bur:
MIDSUMMER NIGHT'S DREAM, Act iii. Scene 2.

Rosalind. O, how full of briers is this working-day world!

Celia. They are but burs, cousin, thrown upon thee in holiday foolery; if we walk not in the trodden paths, our very petticoats will catch them.

Rosalind. I could shake them off my coat; these burs are in my heart.
AS YOU LIKE IT, Act i. Scene 3.

Pandarus. . . . they are burs, I can tell you; they'll stick where they are thrown.

    TROILUS AND CRESSIDA, Act iii. Scene 2.

PLANTAIN.

Costard.   O sir, plantain, a plain plantain; no *l'envoy*, no *l'envoy*, no salve, sir, but a plantain!

    LOVE'S LABOUR'S LOST, Act iii. Scene 1.

Romeo. Your plantain leaf is excellent for that.
Benvolio. For what, I pray thee?
Romeo.   For your broken shin.

    ROMEO AND JULIET, Act i. Scene 2.

CUCKOO FLOWERS.

And cuckoo buds of yellow hue,

    LOVE'S LABOUR'S LOST, Act v. Scene 2.

Cordelia. . . . cuckoo flowers,

    KING LEAR, Act iv. Scene 4.

CLOVER.

Burgundy. . . . green clover,
Wanting the scythe,

    KING HENRY V., Act v. Scene 2.

Tamora. I will enchant the old Andronicus,
With words more sweet, and yet more dangerous
Than baits to fish, or honey stalks to sheep;

    TITUS ANDRONICUS, Act iv. Scene 4.

TREES.

> When the sweet wind did gently kiss the trees,
> And they did make no noise.
> THE MERCHANT OF VENICE, Act v. Scene 1.

OAK.

Prospero. If thou more murmur'st, I will rend an oak.
And peg thee in his knotty entrails, till
Thou hast howl'd away twelve winters.

Prospero. . . . to the dread rattling thunder
Have I given fire, and rifted Jove's stout oak
With his own bolt:
> THE TEMPEST, Act i. Scene 2; Act v. Scene 1.

Mrs. Page. There is an old tale goes, that Herne the hunter,
Sometime a keeper here in Windsor forest,
Doth all the winter-time, at still midnight,
Walk round about an oak,
> MERRY WIVES OF WINDSOR, Act iv. Scene 4.

Benedict. O, she misused me past the endurance of

a block: an oak, but with one green leaf on it, would have answered her;
     MUCH ADO ABOUT NOTHING, Act ii. Scene 1.

 Oliver. Under an old oak, whose boughs were moss'd with age,
And high top bald with dry antiquity,
A wretched ragged man, o'ergrown with hair,
Lay sleeping on his back:
     AS YOU LIKE IT, Act iv. Scene 3.

 Pauline. . . . and will not
. . . . once remove
The root of his opinion, which is rotten,
As ever oak, or stone, was sound.
     A WINTER'S TALE, Act ii. Scene 3.

 Messenger. But Hercules himself must yield to odds;
And many strokes, though with a little axe,
Hew down and fell the hardest-timber'd oak.
     KING HENRY VI., Part III. Act ii. Scene 1.

 Casca.   O Cicero,
I have seen tempests, when the scolding winds
Have rived the knotty oaks;
     JULIUS CÆSAR, Act i. Scene 3.

 Nestor. . . . but when the splitting wind
Makes flexible the knees of knotted oaks,
     TROILUS and CRESSIDA, Act i. Scene 3.

 Iago. She that so young could give out such a seeming,
To seal her father's eyes up, close as oak,—
     OTHELLO, Act iii. Scene 3.

Volumnia. To a cruel war I sent him; from whence he returned, his brows bound with oak.

Volumnia. . . . he comes the third time home with the oaken garland.

2d Guard. The worthy fellow is our general: he is the rock, the oak not to be wind-shaken.

Volumnia. And yet to charge thy sulphur with a bolt
That should but rive an oak.
CORIOLANUS, Act i. Scene 3; Act ii. Scene i.;
Act v. Scenes 2 and 3.

Timon. That numberless upon me stuck, as leaves Do on the oak,
TIMON OF ATHENS, Act iv. Scene 3.

Those thoughts, to me like oaks, to thee like osiers bow'd.
PASSIONATE PILGRIM, v.

CEDAR.

Prospero. . . . and by the spurs pluck'd up
The pine and cedar:
THE TEMPEST, Act v. Scene 1.

Dumain. As upright as the cedar.
LOVE'S LABOUR'S LOST, Act iv. Scene 3.

Warwick. (As on a mountain-top the cedar shows,
That keeps his leaves in spite of any storm,)
KING HENRY VI., Part II. Act v. Scene 1.

Gloucester. Our aiery buildeth in the cedar's top,
And dallies with the wind, and scorns the sun.
King Richard III., Act. i. Scene 3.

Cranmer. . . . he shall flourish,
And, like a mountain cedar, reach his branches
To all the plains about him :———
King Henry VIII., Act v. Scene 5.

Coriolanus. . . . then let the mutinous winds
Strike the proud cedars 'gainst the fiery sun ;
Coriolanus, Act v. Scene 3.

Soothsayer. . . . when from a stately cedar shall be lopped branches, which, being dead many years, shall after revive, be jointed to the old stock, and freshly grow ;

Soothsayer. The lofty cedar, royal Cymbeline,
Personates thee : and thy lopp'd branches point
Thy two sons forth : who, by Belarius stolen,
For many years thought dead, are now revived,
To the majestic cedar join'd ; whose issue
Promises Britain peace and plenty.
Cymbeline, Act v. Scene 5.

Who doth the world so gloriously behold,
The cedar-tops and hills seem burnish'd gold.
Venus and Adonis.

The cedar stoops not to the base shrub's foot,
But low shrubs wither at the cedar's root.
Rape of Lucrece.

ELM.

Adriana. Come, I will fasten on this sleeve of
    thine :
Thou art an elm, my husband, I, a vine ;
            COMEDY OF ERRORS, Act ii. Scene 2.

Titania. . . . . the female ivy so
Enrings the barky fingers of the elm.
            MIDSUMMER NIGHT'S DREAM, Act iv. Scene 1.

Poins. Answer, thou dead elm, answer.
            KING HENRY IV., Part II. Act ii. Scene 4.

SYCAMORE.

Boyet. Under the cool shade of a sycamore,
            LOVE'S LABOUR'S LOST, Act v. Scene 2.

Desdemona. The poor soul sat sighing by a syca-
    more tree,
            OTHELLO, Act iv. Scene 3.

Benvolio. . . . underneath the grove of syca-
    more,
That westward rooteth from this city's side,—
            ROMEO AND JULIET, Act i. Scene 1.

YEW.

Clown. My shroud of white, stuck all with yew,
            TWELFTH NIGHT, Act ii. Scene 4.

Scroop. Thy very beadsmen learn to bend their bows
Of double-fatal yew against thy state;
            KING RICHARD II., Act iii. Scene 2.

Paris. Under yon yew-trees lay thee all along,
Holding thine ear close to the hollow ground;
So shall no foot upon the churchyard tread
  ·   ·   ·   ·   ·   ·   ·   ·
But thou shalt hear it:
            ROMEO AND JULIET, Act v. Scene 3.

Tamora. But straight they told me they would bind me here,
Unto the body of a dismal yew,
            TITUS ANDRONICUS, Act ii. Scene 3.

3d Witch. Gall of goat, and slips of yew,
Sliver'd in the moon's eclipse;
            MACBETH, Act iv. Scene 1.

CYPRESS.

Clown. Come away, come away, death,
And in sad cypress let me be laid;
            TWELFTH NIGHT, Act ii. Scene 4.

Gremio. In ivory coffers I have stuff'd my crowns;
In cypress chests my arras counterpoints,
            TAMING OF THE SHREW, Act ii. Scene 1.

Suffolk. Their sweetest shade a grove of cypress trees!
            KING HENRY VI., Part II. Act iii. Scene 2.

Aufidius. I am attended at the cypress grove:
> CORIOLANUS, Act i. Scene 10.

PINE.

Prospero. . . . she did confine thee,
By help of her more potent ministers,
And in her most unmitigable rage,
Into a cloven pine; within which rift
Imprison'd, thou didst painfully remain
A dozen years;

Prospero. . . . it was mine art,
When I arrived, and heard thee, that made gape
The pine, and let thee out.
> THE TEMPEST, Act i. Scene 2.

1st Lord. Behind the tuft of pines I met them;
> A WINTER'S TALE, Act ii. Scene 1.

Suffolk. Thus droops this lofty pine, and hangs his sprays;
> KING HENRY VI., Part II. Act ii. Scene 3.

Antony. . . . and this pine is bark'd,
That overtopp'd them all.
> ANTONY AND CLEOPATRA, Act iv. Scene 10.

Agamemnon. . . . checks and disasters
Grow in the veins of actions highest rear'd;
As knots, by the conflux of meeting sap,
Infect the sound pine,
> TROILUS AND CRESSIDA, Act i. Scene 3.

Belarius. . . . and yet as rough,
Their royal blood enchafed, as the rudest wind,
That by the top doth take the mountain pine
And make him stoop to the vale.
CYMBELINE, Act iv. Scene 2.

ASPEN.

Hostess. . . . feel, masters, how I shake; look you, I warrant you.
Doll. So you do, hostess.
Hostess. Do I? yea, in very truth, do I, an 'twere an aspen leaf:
KING HENRY IV., Part II. Act ii. Scene 4.

Marcus. Oh! had the monster seen those lily hands
Tremble like aspen-leaves upon a lute,
TITUS ANDRONICUS, Act ii. Scene 5.

BOX.

Maria. Get ye all three into the box-tree:
TWELFTH NIGHT, Act ii. Scene 5.

WILLOW.

Benedict. Come, will you go with me?
Claudio. Whither?
Benedict. Even to the next willow, about your own business, count.

Benedict. . . . I offered him my company to a willow-tree, either to make him a garland, as being forsaken, or to bind him up a rod, as being worthy to be whipped.
MUCH ADO ABOUT NOTHING, Act ii. Scene 1.

Lorenzo. In such a night,
Stood Dido with a willow in her hand
Upon the wild sea-banks, and waft her love
To come again to Carthage.
MERCHANT OF VENICE, Act v. Scene 1.

Bona. Tell him, in hope he'll prove a widower shortly,
I'll wear the willow garland for his sake.
KING HENRY VI., Part III. Act iii. Scene 3.

Desdemona. Sing willow, willow, willow;
Sing all a green willow must be my garland.
OTHELLO, Act iv. Scene 3.

Queen. There is a willow grows aslant a brook,
That shows his hoar leaves in the glassy stream;
HAMLET, Act iv. Scene 7.

PALM.

Rosalind. . . . look here, what I found on a palm-tree:
AS YOU LIKE IT, Act iii. Scene 2.

Cassius. Ye gods, it doth amaze me,
A man of such a feeble temper should
So get the start of the majestic world,
And bear the palm alone.
JULIUS CÆSAR, Act i. Scene 2.

Hamlet. As love between them as the palm should flourish;
HAMLET, Act v. Scene 2.

BAY.

Captain. The bay-trees in our country are all wither'd,
KING RICHARD II., Act ii. Scene 4.

The Vision. Enter, solemnly tripping, one after another, six Personages, clad in white robes, wearing on their heads garlands of bays, and golden vizors on their faces; branches of bays or palm in their hands.
KING HENRY VIII., Act iv. Scene 2.

OLIVE.

Rosalind. . . . if you will know my house, 'Tis at the tuft of olives, here hard by :—
AS YOU LIKE IT, Act iii. Scene 5.

Oliver. Pray you, if you know
Where, in the purlieus of this forest, stands
A sheep-cote, fenced about with olive trees?
AS YOU LIKE IT, Act iv. Scene 3.

Westmoreland. There is not now a rebel's sword unsheathed,
But peace puts forth her olive everywhere.
KING HENRY IV., Part II. Act iv. Scene 4.

Cæsar. The time of universal peace is near:

Prove this a prosperous day, the three-nook'd world
Shall bear the olive freely.
ANTONY AND CLEOPATRA, Act iv. Scene 6.

Incertainties now crown themselves assured,
And peace proclaims olives of endless age.
SONNET CVII.

MYRTLE.

Isabella. Thou rather, with thy sharp and sulphurous
        bolt,
Splitt'st the unwedgeable and gnarled oak,
Than the soft myrtle :
MEASURE FOR MEASURE, Act ii. Scene 2.

Euphronius. . . . 'as petty to his ends,
As is the morn-dew on the myrtle-leaf
To his grand sea.
ANTONY AND CLEOPATRA, Act iii. Scene 10.

.   .   she hasteth to a myrtle grove,
VENUS AND ADONIS.

A cap of flowers, and a kirtle
Embroider'd all with leaves of myrtle.

Sitting in a pleasant shade
Which a grove of myrtles made,
PASSIONATE PILGRIM, xx. and xxi.

HAZEL.

Petrucio.   Kate, like the hazel-twig,
Is straight, and slender ;
TAMING OF THE SHREW, Act ii. Scene 1.

LAUREL.

Clarence. No, Warwick, thou art worthy of the sway,
To whom the heavens, in thy nativity,
Adjudged an olive branch and laurel crown,
As likely to be bless'd in peace, and war;
    King Henry VI., Part III. Act iv. Scene 6.

Ulyss.   How could communities,
Degrees in schools, and brotherhoods in cities,
Peaceful commerce from dividable shores,
The primogenitive and due of birth,
Prerogative of age, crowns, sceptres, laurels,
But by degree, stand in authentic place?
    Troilus and Cressida, Act i. Scene 3.

Titus. Cometh Andronicus, bound with laurel boughs,
    Titus Andronicus, Act i. Scene 2.

ELDER.

Arviragus.   Grow, patience;
And let the stinking elder, grief, untwine
His perishing root with the increasing vine!
    Cymbeline, Act iv. Scene 2.

Saturninus.   Look for thy reward
Among the nettles at the elder tree,

  . . . .

This is the pit, and this the elder tree:
    Titus Andronicus, Act ii. Scene 4.

POMEGRANATE.

Juliet. . . . it is not yet near day:
It was the nightingale, and not the lark,
That pierced the fearful hollow of thine ear;
Nightly she sings on yon pomegranate-tree:
ROMEO AND JULIET, Act iii. Scene 5.

CRAB.

Suffolk. . . . and noble stock
Was graft with crab-tree slip;
KING HENRY VI., Part II. Act iii. Scene 2.

Porter. Fetch me
a dozen crab-tree staves, and strong ones;
KING HENRY VIII., Act v. Scene 4.

Menenius. We have some old crab-trees here at home that will not
Be grafted to your relish.
CORIOLANUS, Act ii. Scene 1.

HOLLY.

Amiens. Heigh ho! sing, heigh ho! unto the green holly:
AS YOU LIKE IT, Act ii. Scene 7.

MISTLETOE.

Tamora. The trees, though summer, yet forlorn and lean,
O'ercome with moss and baleful mistletoe.
TITUS ANDRONICUS, Act ii. Scene 3.

VINE.

Ceres. Vines with clustering bunches growing;
TEMPEST, Act iv. Scene 1.

Mortimer. And pithless arms, like to a wither'd vine
That droops his sapless branches to the ground:
KING HENRY VI., Part I. Act ii. Scene 5.

Cranmer. In her days every man shall eat in safety,
Under his own vine,
KING HENRY VIII., Act v. Scene 5.

Timon. Dry up thy marrows, vines, and plough-torn leas;
TIMON OF ATHENS, Act iv. Scene 3.

Lear.   Now, our joy,
Although our last, not least; to whose young love
The vines of France and milk of Burgundy
Strive to be interess'd;
KING LEAR, Act i. Scene 1.

IVY.

Prospero. . . . he was
The ivy which had hid my princely trunk,
And suck'd my verdure out on't.—
TEMPEST, Act i. Scene 2.

Shepherd. . . . browzing of ivy.
A WINTER'S TALE, Act iii. Scene 3.

Adriana. Usurping ivy, brier, or idle moss;
>> COMEDY OF ERRORS, Act ii. Scene 2.

A belt of straw and ivy buds.
>> PASSIONATE PILGRIM, xvii.

HAWTHORN.

Helena. . . . when hawthorn buds appear
>> MIDSUMMER NIGHT'S DREAM, Act i. Scene. 1.

Rosalind. . . . hangs odes upon hawthorns.
. . all, forsooth, deifying the name of Rosalind:
>> AS YOU LIKE IT, Act iii. Scene 2.

King Henry. Gives not the hawthorn-bush a sweeter shade
To shepherds, looking on their silly sheep,
Than doth a rich embroider'd canopy
To kings, that fear their subjects' treachery?
>> KING HENRY VI., Part III. Act ii. Scene 5.

Edgar. Still through the hawthorn blows the cold wind:
>> KING LEAR. Act iii. Scene 4.

OSIERS.

Nathaniel. Those thoughts to me were oaks, to thee like osiers bow'd.
>> LOVE'S LABOUR'S LOST, Act iv. Scene 2.

Celia. The rank of osiers, by the murmuring stream,
>> AS YOU LIKE IT, Act iv. Scene 3.

BRAMBLES.

> . . . elegies on brambles;
> As You Like It, Act iii. Scene 2.

HIPS. BRIERS. GOSS.

Ariel. . . . so I charm'd their ears,
That, calf-like, they my lowing follow'd, through
Sooth'd briers, sharp furzes, pricking goss, and thorns,
Which enter'd their frail shins:
> Tempest, Act iv. Scene 1.

Rosalind. O, how full of briers is this working-day
world !
> As You Like It, Act i. Scene 3.

Helena. . . . the time will bring on summer
When briers shall have leaves as well as thorns,
And be as sweet as sharp.
> All's Well that Ends Well, Act iv. Scene 4.

Timon. The oaks bear mast, the briers scarlet hips:
> Timon of Athens, Act iv. Scene 3.

LING. HEATH. FURZE. BROOM.

Gonzalo. Now would I give a thousand furlongs of sea for an acre of barren ground ; ling, heath, broom, furze, anything. The wills above be done ! but I would fain die a dry death.
> Tempest, Act i. Scene 1.

FRUITS.

. . . you shall see mine orchard, where in an arbour, we will eat a last year's pippin of my own graffing.
King Henry IV., Part II. Act v. Scene 3.

STRAWBERRY.

Ely. The strawberry grows underneath the nettle;
And wholesome berries thrive and ripen best
Neighbour'd by fruit of baser quality:
King Henry V., Act i. Scene 1.

Gloucester. My lord of Ely, when I was last in Holborn,
I saw good strawberries in your garden there;
I do beseech you send for some of them.

. . . .

Ely. Where is my lord the duke of Gloster?
I have sent for these strawberries.
King Richard III., Act iii. Scene 4.

Iago. Have you not sometimes seen a handkerchief,
Spotted with strawberries, in your wife's hand?
Othello, Act iii. Scene 3.

APRICOTS.

Titania. Be kind and courteous to this gentleman;
Hop in his walks, and gambol in his eyes;
Feed him with apricocks, and dewberries;
   MIDSUMMER NIGHT'S DREAM, Act iii. Scene 1.

Gardener. Go, bind thou up yon' dangling apricocks,
Which, like unruly children, make their sire
Stoop with oppression of their prodigal weight:
   KING RICHARD II., Act iii. Scene 4.

GRAPES.

Titania. Feed him . . . . .
With purple grapes,
   MIDSUMMER NIGHT'S DREAM, Act iii. Scene 1.

Touchstone. The heathen philosopher, when he had a desire to eat a grape, would open his lips when he put it into his mouth; meaning thereby, that grapes were made to eat, and lips to open.
   AS YOU LIKE IT, Act v. Scene 1.

Lafeu. O, will you eat no grapes, my royal fox?
Yes, but you will my noble grapes, an if
My royal fox could reach them:
   ALL'S WELL THAT ENDS WELL, Act ii. Scene 1.

  Come, thou monarch of the vine,
  Plumpy Bacchus, with pink eyne:
  In thy vats our cares be drown'd;
  With thy grapes our hairs be crown'd:
   ANTONY AND CLEOPATRA, Act ii. Scene 7.

Iago. . . . the wine she drinks is made of grapes:
OTHELLO, Act ii. Scene 1.

Menenius. The tartness of his face sours ripe grapes.
CORIOLANUS, Act v. Scene 4.

Even as poor birds, deceived with painted grapes,
VENUS AND ADONIS.

PLUMS.

Cardinal. What, art thou lame?
Simpcox. Ay, . . .
Suffolk. How cam'st thou so?
Simpcox. A fall off of a tree.
Wife. A plum-tree, master.

.   .   .   .   .   .

Gloucester. 'Mass, thou lov'dst plums well, that wouldst venture so.
KING HENRY VI., Part II. Act ii. Scene 1.

Constance. . . . go to it' grandam, child;
Give grandam kingdom, and it' grandam will
Give it a plum,
KING JOHN, Act ii. Scene 1.

The mellow plum doth fall, the green sticks fast.
Or being early pluck'd is sour to taste.
VENUS AND ADONIS.

Like a green plum that hangs upon a tree,
And falls, through wind, before the fall should be.
PASSIONATE PILGRIM, viii.

DAMSONS.

Simpcox. Alas, good master, my wife desired some
damsons,
    King Henry VI., Part II. Act ii. Scene 1.

FIGS.

Titania. Feed him with . . . .
   . . . . green figs,
  Midsummer Night's Dream, Act iii. Scene 1.

Constance. . . . . a fig :
    King John, Act ii. Scene 1.

Charmian. . . . I love long life better than figs.

Guard. Here is a rural fellow
That will not be denied your highness' presence :
He brings you figs.

Dolabella. Who was last with them?
1st Guard. A simple countryman, that brought her
figs.
   Antony and Cleopatra, Act i. Scene 2 ;
   Act v. Scene 2.

PEARS.

Falstaff. I warrant, they would whip me with their
fine wits, till I were as crest-fallen as a dried pear.
   Merry Wives of Windsor, Act iv. Scene 5.

Parolles. 'tis a withered pear ;
  All's Well that Ends Well, Act i. Scene 1.

APPLES.

Sebastian. I think he will carry this island home in his pocket, and give it his son for an apple.
Antonio. And, sowing the kernels of it in the sea, bring forth more islands.
TEMPEST, Act ii. Scene 1.

Evans. . . . I will make an end of my dinner; there's pippins and cheese to come.
MERRY WIVES OF WINDSOR, Act i. Scene 2.

Malvolio. Not yet old enough for a man, nor young enough for a boy; as a squash is before 'tis a peascod, or a codling when 'tis almost an apple:
TWELFTH NIGHT, Act i. Scene 5.

Antonio. An evil soul producing holy witness
Is like a villain with a smiling cheek;
A goodly apple rotten at the heart;
MERCHANT OF VENICE, Act i. Scene 3.

Tranio. He is my father, sir; and, sooth to say,
In countenance somewhat doth resemble you.
Biondello (aside). As much as an apple doth an oyster,

Petrucio. What! up and down, carved like an apple-tart?
TAMING OF THE SHREW, Act iv. Scenes 2 and 3.

1st Drawer. What hast thou brought there? apple-Johns? thou know'st sir John cannot endure an apple-John.
2d Drawer. Thou sayest true. The prince once

set a dish of apple-Johns before him, and told him there were five more sir Johns:

Davy. There is a dish of leather-coats for you.
>   King Henry IV., Part II. Act ii. Scene 4;
>   Act v. Scene 3.

Orleans. Foolish curs! that run winking into the mouth of a Russian bear, and have their heads crushed like rotten apples:
>   King Henry. V., Act iii. Scene 7.

Porter. These are the youths that thunder at a playhouse, and fight for bitten apples:
>   King Henry VIII., Act v. Scene 4.

Fool. Shalt see thy other daughter will use thee kindly; for though she's as like this as a crab's like an apple, yet I can tell what I can tell.
>   King Lear, Act i. Scene 5.

How like Eve's apple doth thy beauty grow,
>   Sonnet XCIII.

QUINCE.

Nurse. They call for dates and quinces in the pastry.
>   Romeo and Juliet, Act iv. Scene 4.

ORANGE.

Beatrice. The count is neither sad, nor sick, nor merry, nor well: but civil count; civil as an orange, and something of that jealous complexion.

Claudio. Give not this rotten orange to your friend ;
   Much Ado about Nothing, Act ii. Scene 1 ;
    Act iv. Scene 1.

Menenius. . . . you wear out a good wholesome forenoon, in hearing a cause between an orange wife and a fosset-seller ;
    Coriolanus, Act ii. Scene 1.

LEMON.

Biron. A lemon.
    Love's Labour's Lost, Act v. Scene 2.

CHERRIES.

Demetrius.   O, how ripe in show
Thy lips, those kissing cherries, tempting grow!
.   .   .   .

Helena.   So we grew together,
Like to a double cherry.
    Midsummer Night's Dream, Act iii. Scene 2.

Constance. A cherry,
    King John, Act ii. Scene 1.

Lady. . . . 'tis as like you
As cherry is to cherry.
    King Henry VIII., Act v. Scene 1.

Gower. . and with her neeld composes
Nature's own shape, of bud, bird, branch, or berry ;
That even her art sisters the natural roses ;
Her inkle, silk, twin with the rubied cherry :
    Pericles, Act v.

. . . . . ripe red cherries;
VENUS AND ADONIS.

GOOSEBERRY.

Falstaff. . . . not worth a gooseberry.
KING HENRY IV., Part II. Act i. Scene 2.

MULBERRY.

Titania. Feed him with . . . . .
. . . . . . mulberries;
MIDSUMMER NIGHT'S DREAM, Act iii. Scene 1.

Volumnia. . . . . correcting thy stout heart,
Now humble, as the ripest mulberry
That will not hold the handling:
CORIOLANUS, Act iii. Scene 2.

When he was by, the birds such pleasure took,
That some would sing, some other in their bills
Would bring him mulberries,
VENUS AND ADONIS.

MEDLARS.

Touchstone. Truly, the tree yields bad fruit.
Rosalind. I'll graff it with you, and then I shall graff it with a medlar: then it will be the earliest fruit in the country: for you'll be rotten ere you be half ripe, and that's the right virtue of the medlar.
AS YOU LIKE IT, Act iii. Scene 2.

Apemantus. There's a medlar for thee, eat it.
Timon. On what I hate I feed not.
Apemantus. Dost hate a medlar?
Timon. Ay, though it look like thee.
     Timon of Athens, Act iv. Scene 3.

Mercutio. Now will he sit under a medlar tree,
     Romeo and Juliet, Act ii. Scene 1.

BLACKBERRY.

Falstaff. . . . if reasons were as plenty as blackberries I would give no man a reason upon compulsion,
     King Henry IV., Part I. Act ii. Scene 4.

Thersites. . . . —— that stale old mouse-eaten dry cheese, Nestor, and that same dog-fox, Ulysses,—is not proved worth a blackberry.—
     Triolus and Cressida, Act v. Scene 4.

CRAB.

When roasted crabs hiss in the bowl,
     Love's Labour's Lost, Act v. Scene 2.

Petrucio. Nay, come, Kate, come; you must not look so sour.
Katharine. It is my fashion, when I see a crab.
Petrucio. Why, here's no crab; and therefore look not sour.
     Taming of the Shrew, Act ii. Scene 1.

Fool. She will taste as like this as a crab does to a crab.
KING LEAR, Act i. Scene 5.

BILBERRY.

Pistol. Cricket, to Windsor chimneys shalt thou
  leap :
Where fires thou find'st unraked and hearths unswept,
There pinch the maids as blue as bilberry ;
MERRY WIVES OF WINDSOR, Act v. Scene 5.

DEWBERRY.

Titania. Feed him with   . . .   dewberries ;
MIDSUMMER NIGHT'S DREAM, Act iii. Scene 1.

NUTS.

Sweetest nut hath sourest rind,
Such a nut is Rosalind.
AS YOU LIKE IT, Act iii. Scene 2.

Thersites. Hector shall have a great catch if he knock out either of your brains ; 'a were as good crack a fusty nut with no kernel.
TROILUS AND CRESSIDA, Act ii. Scene 1.

Hamlet. . . . such officers do the king best service in the end : he keeps them, like an ape doth nuts, in the corner of his jaw ; first mouthed, to be last swallowed :
HAMLET, Act iv. Scene 2.

Mercutio.   Thou wilt quarrel with a man for cracking nuts,
ROMEO AND JULIET, Act iii. Scene 1.

WALNUT.

Ford. . . . As jealous as Ford, that searched a hollow walnut
MERRY WIVES OF WINDSOR, Act iv. Scene 2.

Haberdasher. Here is the cap your worship did bespeak.
Petrucio. Why, this was moulded on a porringer; A velvet dish;— . . .
Why, 'tis a cockle, or a walnut-shell,
TAMING OF THE SHREW, Act iv. Scene 3.

HAZEL-NUT.

Petrucio. . . . as brown in hue, As hazel-nuts, and sweeter than the kernels.
TAMING OF THE SHREW, Act ii. Scene 1.

Mercutio. Her chariot is an empty hazel-nut,
ROMEO AND JULIET, Act i. Scene 4.

CHESTNUT.

Petrucio. And do you tell me of a woman's tongue, That gives not half so great a blow to th' ear As will a chestnut in a farmer's fire?
TAMING OF THE SHREW, Act i. Scene 2.

1st Witch. A sailor's wife had chestnuts in her lap, And mounch'd, and mounch'd, and mounch'd :
MACBETH, Act i. Scene 3.

PRUNES.

Slender. . . . I bruised my shin the other day with playing at sword and dagger with a master of fence, three veneys for a dish of stewed prunes ;
MERRY WIVES OF WINDSOR, Act i. Scene 1.

Pompey. . . . you being then, if you be remembered, cracking the stones of the foresaid prunes.
MEASURE FOR MEASURE, Act ii. Scene 1.

Clown. . . . *four pound of prunes,*
A WINTER'S TALE, Act iv. Scene 2.

Doll. He a captain ! . . . He lives upon mouldy stewed prunes and dried cakes.
KING HENRY IV., Part II. Act ii. Scene 4.

DATES.

Clown. . . . *Dates,*—none ; that's out of my note :
A WINTER'S TALE, Act iv. Scene 2.

CURRANTS.

Clown. . . . *Three pound of sugar ; five pound of currants ;*
A WINTER'S TALE, Act iv. Scene 2.

RAISINS.

Clown. . . . *four pound of prunes*, and as many of *raisins o' the sun*.
     A WINTER'S TALE, Act iv. Scene 2.

ACORN.

Prospero. . . . wither'd roots, and husks
Wherein the acorn cradled.
     THE TEMPEST, Act i. Scene 2.

Puck. . . . all their elves, for fear,
Creep into acorn-cups, and hide them there.
    MIDSUMMER NIGHT'S DREAM, Act ii. Scene 1.

Celia. . . . I found him under a tree, like a dropped acorn.
     AS YOU LIKE IT, Act iii. Scene 2.

BERRIES.

Caliban.   When thou camest first,
Thou strok'dst me, and mad'st much of me; would'st give me
Water with berries in't;

Caliban. I'll show thee the best springs; I'll pluck thee berries;
   THE TEMPEST, Act i. Scene 2; Act ii. Scene 2.

1st Bandit. We cannot live on grass, on berries, water,
As beasts, and birds, and fishes.
TIMON OF ATHENS, Act iv. Scene 3.

Cæsar. . . . thy palate then did deign
The roughest berry on the rudest hedge;
ANTONY AND CLEOPATRA, Act i. Scene 4.

VEGETABLES.

Guiderius. But his neat cookery! He cut our roots in characters:
CYMBELINE, Act iv. Scene 2.

POTATO.

Falstaff. . . . Let the sky rain potatoes;
MERRY WIVES OF WINDSOR, Act v. Scene 5.

CABBAGE.

Evans. Pauca verba, sir John, goot worts.
Falstaff. Good worts! good cabbage.—
MERRY WIVES OF WINDSOR, Act i. Scene 1.

TURNIPS.

Anne. Alas, I had rather be set quick i' the earth,
And bowl'd to death with turnips.
MERRY WIVES OF WINDSOR, Act iii. Scene 4.

RADISH.

Prince Henry. What, fought ye with them all?
Falstaff. All? I know not what ye call all; but

if I fought not with fifty of them I am a bunch of radish :

   King Henry IV., Part I. Act ii. Scene 4.

PARSLEY.

Biondello. . . . I knew a wench married in an afternoon as she went to the garden for parsley to stuff a rabbit ;

   Taming of the Shrew, Act iv. Scene 4.

PEASCOD.

Touchstone. . . . I remember, when I was in love, I broke my sword upon a stone, and bid him take that for coming a'night to Jane Smile : . . . and I remember the wooing of a peascod instead of her; from whom I took two cods, and, giving her them again, said, with weeping tears, *Wear these for my sake.*

   As You Like It, Act ii. Scene 4.

Malvolio. . . . as a squash is before 'tis a peascod,

   Twelfth Night, Act i. Scene 5.

Hostess. Well, fare thee well : I have known thee these twenty-nine years, come peascod time ;

   King Henry IV., Part II. Act ii. Scene 4.

Fool (pointing to Lear). That's a sheal'd peascod.

   King Lear, Act i. Scene 4.

LETTUCE.

Iago. . . . if we will plant nettles, or sow lettuce;

OTHELLO, Act i. Scene 3.

MUSHROOM.

Prospero. . . . and you, whose pastime
Is to make midnight mushrooms;

TEMPEST, Act v. Scene 1.

SAMPHIRE.

Edgar. . . . half way down
Hangs one that gathers samphire; dreadful trade!

KING LEAR, Act iv. Scene 6.

ONION.

Bottom. And, most dear actors, eat no onions,

MIDSUMMER NIGHT'S DREAM, Act iv. Scene 2.

Lord. And if the boy have not a woman's gift,
To rain a shower of commanded tears,
An onion will do well for such a shift;
Which in a napkin being close convey'd
Shall in despite enforce a watery eye.

TAMING OF THE SHREW (Induction), Scene 1.

Lafeu. Mine eyes smell onions, I shall weep anon:—

ALL'S WELL THAT ENDS WELL, Act v. Scene 3.

Enobarbus. —and, indeed, the tears live in an onion that should water this sorrow.

Enobarbus. Look, they weep ;
And I, an ass, am onion-eyed ;
>> ANTONY AND CLEOPATRA, Act. i. Scene 2 ;
>> Act iv. Scene 2.

GARLIC.

Bottom. And, most dear actors, eat no onions,
Nor garlic, for we are to utter sweet breath ;
>> MIDSUMMER NIGHT'S DREAM, Act iv. Scene 2.

Dorcas. Mopsa must be your mistress : marry,
  garlic,
To mend her kissing with.
>> A WINTER'S TALE, Act iv. Scene 3.

Hotspur. —I had rather live
With cheese and garlic in a windmill, far,
Than feed on cates, and have him talk to me,
>> KING HENRY IV., Part I. Act iii. Scene 1.

Menenius. You have made good work,
You, and your apron-men ; you that stood so much
Upon the voice of occupation, and
The breath of garlic-eaters !
>> CORIOLANUS, Act iv. Scene 6.

LEEK.

Thisbe. His eyes were green as leeks,
>> MIDSUMMER NIGHT'S DREAM, Act v. Scene 2.

Pistol. Knowest thou Fluellen ?
King Henry. Yes.
Pistol. Tell him, I'll knock his leek about his pate,
Upon Saint Davy's day.

Fluellen. . . . if your majesties is remembered of it, the Welshmen did goot service in a garden where leeks did grow, wearing leeks in their Monmouth caps; which, your majesty know, to this hour is an honourable padge of the service; and, I do believe, your majesty takes no scorn to wear the leek upon Saint Tavy's day.

Fluellen. . . . he is come to me, and prings me pread and salt yesterday, look you, and bid me eat my leek: . . . but I will be so pold as to wear it in my cap till I see him once again, and then I will tell him a little piece of my desires.

Pistol. . . . I am qualmish at the smell of leek.
Fluellen. I peseech you . . . to eat, look you, this leek; . . . if you can mock a leek, you can eat a leek.
I say, I will make him eat some part of my leek, . .
Pistol. Must I bite?
Fluellen. Yes, certainly; . . .
Pistol. By this leek, I will most horribly revenge; I eat—and eat—I swear.
Fluellen. Eat, I pray you: will you have some more sauce to your leek? there is not enough leek to swear by.

. . . .

When you take occasions to see leeks hereafter, I pray you, mock at 'em; that is all.
Pistol. Good.
Fluellen. Ay, leeks is goot :—
KING HENRY V., Act iv. Scenes 1 and 7;
Act v. Scene 1.

HERBS.

Friar. O mickle is the powerful grace, that lies
In plants, herbs, stones, and their true qualities:
        ROMEO AND JULIET, Act ii. Scene 3.

MINT.

Dumain.   That mint.
        LOVE'S LABOUR'S LOST, Act v. Scene 2.

Perdita.   —Here's flowers for you;
. . . . mints,
        A WINTER'S TALE, Act iv. Scene 3.

SAVORY.

Perdita. . . . . . savory,
        A WINTER'S TALE, Act iv. Scene 3.

MARJORAM.

Perdita. . . . . . marjoram;
        A WINTER'S TALE, Act iv. Scene 3.

Lafeu. . . . we may pick a thousand sallets, ere we light on such another herb.

Clown. Indeed, sir, she was the sweet marjoram of the sallet, or, rather, the herb of grace.

Lafeu. They are not sallet-herbs, you knave, they are nose-herbs.

ALL'S WELL THAT ENDS WELL, Act iv. Scene 5.

King Lear. —Give the word.
Edgar. Sweet marjoram.

KING LEAR, Act iv. Scene 6.

And buds of marjoram had stolen thy hair:

SONNET XCIX.

RUE.

Perdita. Reverend sirs,
For you there's . . . rue;

A WINTER'S TALE, Act iv. Scene 3.

Gardener. . . . here, in this place,
I'll set a bank of rue, sour herb of grace:
Rue, even for ruth, here shortly shall be seen,
In the remembrance of a weeping queen.

KING RICHARD II., Act iii. Scene 4.

Ophelia. —there's rue for you; and here's some for me:—we may call it herb-grace o' Sundays:—oh, you must wear your rue with a difference.——

HAMLET, Act iv. Scene 5.

ROSEMARY.

Perdita. For you there's rosemary, and rue ; these keep
Seeming, and savour, all the winter long :
        A WINTER'S TALE, Act iv. Scene 3.

Ophelia. There's rosemary, that's for remembrance ;
        HAMLET, Act iv. Scene 5.

Nurse. . . . Doth not rosemary and Romeo begin both with a letter ?

Friar. Dry up your tears, and stick your rosemary
On this fair corse ;
        ROMEO AND JULIET, Act ii. Scene 4 ;
        Act iv. Scene 5.

Edgar. . . . sprigs of rosemary ;
        KING LEAR, Act ii. Scene 3.

FENNEL.

Doll. Why doth the prince love him so then ?
Falstaff. Because their legs are both of a bigness : and he plays at quoits well ; and eats conger and fennel ;
        KING HENRY IV., Part II. Act ii. Scene 4.

Ophelia. There's fennel for you,
        HAMLET, Act iv. Scene 5.

HYSSOP AND THYME.

Iago. . . . Our bodies are our gardens; to which our wills are gardeners: so that if we will plant nettles, or sow lettuce; set hyssop, and weed up thyme; supply it with one gender of herbs, or distract it with many; either to have it sterile with idleness, or manured with industry; why, the power and corrigible authority of this lies in our wills.

OTHELLO, Act i. Scene 3.

Oberon. I know a bank where the wild thyme blows,
MIDSUMMER'S NIGHT'S DREAM, Act ii. Scene 1.

BALM.

Anne. The several chairs of order look you scour
With juice of balm, and every precious flower:
MERRY WIVES OF WINDSOR, Act v. Scene 5.

BURNET.

Burgundy. The freckled cowslip, burnet, and green clover,
KING HENRY V., Act v. Scene 2.

SPICES AND MEDICINES.

Clown.   I must go buy spices.
        A WINTER'S TALE, Act iv. Scene 2.

NUTMEG.

Armado. The armipotent Mars, of lances the almighty,
Gave Hector a gift,——
Dumain. A gilt nutmeg.
        LOVE'S LABOUR'S LOST, Act v. Scene 2.

Clown. . . . nutmegs, seven ;
        A WINTER'S TALE, Act iv. Scene 2.

Orleans. He's of the colour of the nutmeg.
        KING HENRY V., Act iii. Scene 7.

CLOVE.

Longaville. Stuck with cloves.
        LOVE'S LABOUR'S LOST, Act v. Scene 2.

GINGER.

Dauphin. And of the heat of the ginger.
        KING HENRY V., Act iii. Scene 7.

Costard. An I had but one penny in the world, thou shouldst have it to buy gingerbread:
>> Love's Labour's Lost, Act v. Scene 1.

Clown. . . . *a race or two of ginger;* but that I may beg;—
>> A Winter's Tale, Act iv. Scene 2.

2d Carrier. I have a gammon of bacon, and two razes of ginger, to be delivered as far as Charing Cross.
>> King Henry IV., Part I. Act ii. Scene 1.

MACE.

Clown. . . *mace,—dates,*—none;
>> A Winter's Tale, Act ii. Scene 2.

SAFFRON.

Clown. I must have *saffron*, to colour the warden pies;
>> A Winter's Tale, Act iv. Scene 2.

PEPPERCORN.

Falstaff. . . . An I have not forgotten what the inside of a church is made of, I am a peppercorn,
>> King Henry IV., Part I. Act iii. Scene 3.

PEPPER.

Ford. . . . he cannot creep into a halfpenny purse, nor into a pepper box;
    THE MERRY WIVES OF WINDSOR, Act iii. Scene 5.

Sir Andrew. Here's the challenge, read it; I warrant there's vinegar and pepper in't.
    TWELFTH NIGHT, Act iii. Scene 4.

MUSTARD.

Bottom. . . . Your name, I beseech you, sir?
Mustard. Mustard-seed.
Bottom. Good master Mustard-seed, I know your patience well: that same cowardly, giant like ox-beef hath devoured many a gentleman of your house: I promise you, your kindred hath made my eyes water ere now. I desire you more acquaintance, good master Mustard-seed.
    MIDSUMMER NIGHT'S DREAM, Act iii. Scene 1.

Touchstone. . . . by mine honour; . . .
Rosalind. Where learned you that oath, fool?
Touchstone. Of a certain knight, that swore by his honour they were good pancakes, and swore by his honour the mustard was naught; now, I'll stand to it, the pancakes were naught, and the mustard was good; and yet was not the knight forsworn.
Celia. How prove you that, in the great heap of your knowledge?
Rosalind. Ay, marry; now unmuzzle your wisdom.

Touchstone. Stand you both forth now: stroke your chins, and swear by your beards that I am a knave.

Celia. By our beards, if we had them, thou art.

Touchstone. By my knavery, if I had it, then I were: but if you swear by that that is not, you are not forsworn: no more was this knight, swearing by his honour, for he never had any; or if he had, he had sworn it away before ever he saw those pancakes or that mustard.

As You Like It, Act i. Scene 2.

Grumio. . . . What say you to a piece of beef and mustard?

Katharine. A dish that I do love to feed upon.

Grumio. Ay, but the mustard is too hot a little.

Katharine. Why, then the beef, and let the mustard rest.

Grumio. Nay, then I will not; you shall have the mustard,

Or else you get no beef of Grumio.

Katharine. Then both, or one, or anything thou wilt.

Grumio. Why, then the mustard without the beef.

Taming of the Shrew, Act iv. Scene 3.

Falstaff. . . . his wit is as thick as Tewksbury mustard;

King Henry IV., Part II. Act ii. Scene 4.

RHUBARB AND SENNA.

Macbeth. What rhubarb, senna, or what purgative drug,
Would scour these English hence?

Macbeth, Act v. Scene 3.

POPPY AND MANDRAGORA.

Iago. Not poppy, nor mandragora,
Nor all the drowsy syrups of the world,
Shall ever medicine thee to that sweet sleep
Which thou ow'dst yesterday.
OTHELLO, Act iii. Scene 3.

> *Cleopatra.* . . . Give me to drink mandragora.
> *Charmian.* Why, madam?
> *Cleopatra.* That I might sleep out this great gap of time
> ANTONY AND CLEOPATRA, Act i. Scene 5.

HEBENON.

Ghost. Sleeping within mine orchard,
My custom always in the afternoon,
Upon my secure hour thy uncle stole,
With juice of cursed hebenon in a vial,
And in the porches of mine ears did pour
The leperous distilment;
HAMLET, Act i. Scene 5.

GUM.

Othello. . . . of one, whose subdued eyes,
Albeit unusèd to the melting mood,
Drop tears as fast as the Arabian trees
Their med'cinable gum.
OTHELLO, Act v. Scene 2.

Poet. . . . .
Our poesy is as a gum, which oozes
From whence 'tis nourish'd:
TIMON OF ATHENS, Act i. Scene 1.

Hamlet. . . plum tree gum; . . .
HAMLET, Act ii. Scene 2.

CAMOMILE.

Falstaff. . . . for though the camomile, the more it is trodden the faster it grows, yet youth, the more it is wasted the sooner it wears.
KING HENRY IV., Part I. Act ii. Scene 4.

ACONITUM.

King Henry. . . . as strong
As aconitum,
KING HENRY IV., Part II. Act iv. Scene 4.

COLOQUINTIDA. LOCUSTS.

Iago. . . . the food that to him now is as luscious as locusts, shall be to him shortly as bitter as coloquintida.
OTHELLO, Act i. Scene 3.

GRAIN.

You sun-burn'd sicklemen, of August weary,
Come hither from the furrow, and be merry;
      THE TEMPEST, Act iv. Scene 1.

CORN.

Gonzalo. No use of metal, corn, or wine, or oil:
      THE TEMPEST, Act ii. Scene 1.

Duke.   Come, let us go:
Our corn's to reap, for yet our tithe's to sow.
      MEASURE FOR MEASURE, Act iv. Scene 1.

King Richard. . . . my tender-hearted cousin!—
We'll make foul weather with despisèd tears;
Our sighs, and they, shall lodge the summer corn,
And make a dearth in this revolting land.
      KING RICHARD II., Act iii. Scene 3.

La Pucelle. Talk like the vulgar sort of market-men
That come to gather money for their corn.

Guard (within). Qui est la ?
Pucelle. Paisans, pauvres gens de France :
Poor market-folks, that come to sell their corn.

Pucelle. Good morrow, gallants! want ye corn for bread?
KING HENRY VI., Part I. Act iii. Scene 2.

Duchess. Why droops my lord, like over-ripen'd corn,
Hanging the head at Ceres' plenteous load?
KING HENRY VI., Part II. Act i. Scene 2.

Cranmer (kneeling). I humbly thank your highness;
And am right glad to catch this good occasion
Most thoroughly to be winnow'd, where my chaff
And corn shall fly asunder :
KING HENRY VIII., Act v. Scene 1.

1st Citizen. . . . we'll have corn at our own price.

Marcius. . . . What's their seeking?
Menenius. For corn at their own rates; whereof, they say,
The city is well stored.

Marcius. . . . .
They said they were an-hungry; sighed forth proverbs,—
That hunger broke stone walls; . . . . .
That meat was made for mouths; that the gods sent not
Corn for the rich men only :—

1st Citizen. . . . we stood up about the corn,

Brutus. . . . . The people cry you mock'd them ; and, of late,
When corn was given them gratis, you repined ;

Coriolanus. Tell me of corn !
This was my speech, and I will speak't again ;—
   CORIOLANUS, Act i. Scene 1. ; Act ii. Scene 3 ;
    Act iii. Scene 1.

Pericles. And these our ships . . .
 . . . .
Are stored with corn to make your needy bread,
And give them life, whom hunger starved half-dead.
    PERICLES, Act i. Scene 4.

Macbeth. Though bladed corn be lodged, and trees blown down ;
    MACBETH, Act iv. Scene 1.

WHEAT.

Iris. Ceres, most bounteous lady, thy rich leas of wheat,
    TEMPEST, Act iv. Scene 1.

Helena. When wheat is green, when hawthorn buds appear.
   MIDSUMMER NIGHT'S DREAM, Act i. Scene 1.

Davy. . . . shall we sow the head-land with wheat?
Shallow. With red wheat, Davy.
   KING HENRY IV., Part II. Act v. Scene 1.

Pompey. . . . I must
Rid all the sea of pirates; then, to send
Measures of wheat to Rome:
ANTONY AND CLEOPATRA, Act ii. Scene 6.

Pandarus. . . . . . .
He that will have a cake out of the wheat must needs
tarry the grinding.
TROILUS AND CRESSIDA, Act i. Scene 1.

Hamlet. . . . . . .
As peace should still her wheaten garland wear,
HAMLET, Act v. Scene 2.

Edgar. This is the foul fiend Flibbertigibbet: he
begins at curfew, and walks till the first cock; he
. . . . mildews the white wheat,
KING LEAR, Act iii. Scene 4.

BARLEY.

Iris. Ceres, most bounteous lady, thy rich leas
of . . . barley,
TEMPEST, Act iv. Scene 1.

OATS.

Iris. . . . oats, . . .
TEMPEST, Act iv. Scene 1.

Titania. Or say, sweet love, what thou desir'st to eat.
Bottom. Truly, a peck of provender: I could
munch your good dry oats.
MIDSUMMER NIGHT'S DREAM, Act iv. Scene 1.

Petrucio. Grumio, my horse.
Grumio. Ay, sir, they be ready ; the oats have eaten the horses.
   TAMING OF THE SHREW, Act iii. Scene 2.

2d Carrier. . . . this house is turned upside down since Robin ostler died.
1st Carrier. Poor fellow ! never joyed since the price of oats rose ; it was the death of him.
   KING HENRY IV., Part I. Act ii. Scene 1.

Officer. I cannot draw a cart, nor eat dried oats ;
   KING LEAR, Act v. Scene 3.

RYE.

Iris. . . . rye, . . . your rye straw hats put on,
   TEMPEST, Act iv. Scene 1.

VETCHES.

Iris. . . . vetches, . . .
   TEMPEST, Act iv. Scene 1.

PEAS.

Iris. . . . . pease ;
   TEMPEST, Act iv. Scene 1.

Bottom. I had rather have a handful, or two, of dried peas.
   MIDSUMMER NIGHT'S DREAM, Act iv. Scene 1.

Biron. This fellow pecks up wit, as pigeons peas,
>> LOVE'S LABOUR'S LOST, Act v. Scene 2.

BEANS.

2d Carrier. Peas and beans are as dank here as a dog,
>> KING HENRY IV., Part I. Act ii. Scene 1.

FLAX.

Sir Toby. Then hads't thou had an excellent head of hair.
. . . it hangs like flax on a distaff;
>> TWELFTH NIGHT, Act i. Scene 3.

3d Servant. . . . I'll fetch some flax, and whites of eggs,
To apply to his bleeding face.
>> KING LEAR, Act iii. Scene 7.

RICE.

Clown. . . . rice.—What will this sister of mine do with rice?
>> WINTER'S TALE, Act iv. Scene 2.

HEMP.

Hostess. Good people, bring a rescue. Thou wilt not? thou wilt not? do, do, thou rogue! do, thou hemp-seed!
>> KING HENRY IV., Part II. Act ii. Scene 1.

BIRDS.

Tamora. The birds chant melody on every bush;
TITUS ANDRONICUS, Act ii. Scene 3.

OSTRICH.

Cade. . . . I'll make thee eat iron like an ostrich, and swallow my sword like a great pin, ere thou and I part.
KING HENRY VI., Part II. Act iv. Scene 10.

Enobarbus. . . . To be furious,
Is to be frighted out of fear: and in that mood,
The dove will peck the estridge;
ANTONY AND CLEOPATRA, Act iii. Scene 11.

PEACOCK.

Iris. . . . her peacocks fly amain:
TEMPEST, Act iv. Scene 1.

Dromio S. Fly pride, says the peacock:
COMEDY OF ERRORS, Act iv. Scene 3.

Williams. . . . you may as well go about to turn the sun to ice, with fanning in his face with a peacock's feather.
KING HENRY V., Act iv. Scene 1.

La Pucelle. Let frantic Talbot triumph for a while,
And like a peacock sweep along his tail;
We'll pull his plumes, and take away his train,
If Dauphin and the rest will be but ruled.
KING HENRY VI., Part I. Act iii. Scene 3.

Thersites. Why, he stalks up and down like a peacock,—a stride, and a stand:
TROILUS AND CRESSIDA, Act iii. Scene 3.

PARROT.

Dromio E. Mistress, *respice finem*, respect your end; or rather the prophecy, like the parrot, *Beware the rope's end.*
COMEDY OF ERRORS, Act iv. Scene 4.

Benedict. Well, you are a rare parrot-teacher.
Beatrice. A bird of my tongue is better than a beast of yours.
MUCH ADO ABOUT NOTHING, Act i. Scene 1.

Rosalind. . . . I will be more jealous of thee than a Barbary cock-pigeon over his hen; more clamorous than a parrot against rain;
AS YOU LIKE IT, Act iv. Scene 1.

Salarino. Nature hath framed strange fellows in her time:

Some that will evermore peep through their eyes,
And laugh, like parrots, at a bagpiper :

Lorenzo. I think, the best grace of wit will shortly turn into silence ; and discourse grow commendable in none only but parrots.
>> MERCHANT OF VENICE, Act i. Scene 1 ;
>> Act iii. Scene 5.

Prince Henry. That ever this fellow should have fewer words than a parrot, and yet the son of a woman !
>> KING HENRY IV., Part i. Act ii. Scene 4.

EAGLE.

Bast. . . . know, the gallant monarch is in arms ;
And, like an eagle o'er his aiery, towers,
To souse annoyance that comes near his nest.—
>> KING JOHN, Act v. Scene 2.

York. Yet looks he like a king ; behold, his eye,
As bright as is the eagle's, lightens forth
Controlling majesty :
>> KING RICHARD II., Act iii. Scene 3.

Falstaff. . . . when I was about thy years, Hal, I was not an eagle's talon in the waist ;
>> KING HENRY IV., Part I. Act ii. Scene 4.

King Henry. Poor queen ! how love to me, and to her son,
Hath made her break out into terms of rage !

Revenged may she be on that hateful duke;
Whose haughty spirit, winged with desire,
Will cost my crown, and, like an empty eagle,
Tire on the flesh of me and of my son!
KING HENRY VI., Part III. Act i. Scene 1.

Cassius. Coming from Sardis, on our former ensign
Two mighty eagles fell; and there they perch'd,
Gorging and feeding from our soldiers' hands,
Who to Philippi here consorted us;
This morning are they fled away, and gone;
And in their steads do ravens, crows, and kites,
Fly o'er our heads, and downward look on us,
As we were sickly prey;
JULIUS CÆSAR, Act v. Scene 1.

Mecænas. Eight wild boars roasted whole at a breakfast, and but twelve persons there: is this true?

Enobarbus. This was but as a fly by an eagle: we had much more monstrous matter of feasts, which worthily deserved noting.
ANTONY AND CLEOPATRA, Act ii. Scene 2.

Coriolanus. If you have writ your annals true, 'tis there,
That like an eagle in a dove-cote, I
Flutter'd your Volscians in Corioli:
CORIOLANUS, Act v. Scene 6.

Poet. . . . no levell'd malice
Infects one comma in the course I hold;
But flies an eagle flight, bold, and forth on,
Leaving no track behind.
TIMON OF ATHENS, Act i. Scene 1.

Nurse. . . . an eagle, madam,
Hath not so green, so quick, so fair an eye,
As Paris hath.
ROMEO AND JULIET, Act iii. Scene 5.

Cleon.   Thou art like the harpy,
Which, to betray, dost, with thine angel's face,
Seize with thine eagle's talons.
PERICLES, Act iv. Scene 4.

Soothsayer. I saw Jove's bird, the Roman eagle, wing'd
From the spungy south to this part of the west,
There vanish'd in the sunbeams: which portends,
(Unless my sins abuse my divination),
Success to the Roman host.
CYMBELINE, Act iv. Scene 2.

VULTURE.

Pistol. Let vultures vile seize on his lungs also!
KING HENRY IV., Part II. Act v. Scene 3.

Lear. . . . O Regan, she hath tied
Sharp-tooth'd unkindness, like a vulture, here,—
[*Points to his heart.*
KING LEAR, Act ii. Scene 4.

FALCON.

Petrucio. My falcon now is sharp, and passing empty:
TAMING OF THE SHREW, Act iv. Scene 1.

Isabella. . . . This outward-sainted deputy,—
Whose settled visage and deliberate word
Nips youth i' the head, and follies doth emmew,
As falcon doth the fowl,—
    MEASURE FOR MEASURE, Act iii. Scene 1.

Bolingbroke. As confident as is the falcon's flight
Against a bird, do I with Mowbray fight.
    KING RICHARD II., Act. i. Scene 3.

King Henry. But what a point, my lord, your falcon made,
And what a pitch she flew above the rest !——
    KING HENRY VI., Part II. Act ii. Scene 1.

Pandarus. The falcon as the tercel, for all the ducks i' the river :
    TROILUS AND CRESSIDA, Act. iii. Scene 2.

Old Man. . . . On Tuesday last,
A falcon, towering in her pride of place,
Was by a mousing owl hawk'd at and kill'd.
    MACBETH, Act ii. Scene 4.

HAWK.

Page. . . . I do invite you to-morrow morning to my house to breakfast : after, we'll a-birding together ; I have a fine hawk for the bush.
    MERRY WIVES OF WINDSOR, Act iii. Scene 3.

Lord. Dost thou love hawking ? thou hast hawks will soar
Above the morning lark :
    TAMING OF THE SHREW, Induction, Scene 2.

Dauphin. . . . *le cheval volant,* the Pegasus, *qui a les narines de feu!* When I bestride him I soar, I am a hawk :
KING HENRY V., Act iii. Scene 7.

Warwick. Between two hawks, which flies the higher pitch,
KING HENRY VI., Part I. Act ii. Scene 4.

Suffolk. No marvel, an it like your majesty,
My lord protector's hawks do tower so well ;
They know their master loves to be aloft,
And bears his thoughts above his falcon's pitch.
KING HENRY VI., Part II. Act ii. Scene 1.

Some glory in their birth, some in their skill,
. . . .
Some in their hawks and hounds, some in their horse ;
SONNET XCI.

HAGGARD (AN UNTAMED HAWK).

Hero. . . . . .
No, truly, Ursula, she is too disdainful ;
I know, her spirits are as coy and wild
As haggards of the rock.
MUCH ADO ABOUT NOTHING, Act iii. Scene 1.

Viola. This fellow is wise enough to play the fool ;
And to do that well craves a kind of wit :
He must observe their mood on whom he jests,
The quality of persons, and the time ;
And, like the haggard, check at every feather
That comes before his eye.
TWELFTH NIGHT, Act iii. Scene 1.

G

Petrucio. Another way I have to man my haggard,
To make her come, and know her keeper's call,

Hortensio. I will be married to a wealthy widow
Ere three days pass; which hath as long loved me,
As I have loved this proud disdainful haggard:
TAMING OF THE SHREW, Act iv. Scenes 1 and 2.

Othello. . . . If I do prove her haggard,
Though that her jesses were my dear heart-strings,
I'd whistle her off, and let her down the wind,
To prey at fortune.
OTHELLO, Act iii. Scene 3.

STANNYEL (THE COMMON STONE HAWK).

Sir Toby. And with what wing the stannyel checks at it!
TWELFTH NIGHT, Act ii. Scene 5.

EYAS (A YOUNG HAWK).

Rosencrantz. . . . but there is, sir, an aiery of children, little eyases, that cry out on the top of question,
HAMLET, Act ii. Scene 2.

EYAS-MUSKET.

Mrs. Page. Here comes little Robin.
Mrs. Ford. How now, my eyas-musket? what news with you?
THE MERRY WIVES OF WINDSOR, Act iii. Scene 3.

COYSTRIL (A Coward Hawk).

Sir Toby. . . . He's a coward and a coystril,
TWELFTH NIGHT, Act i. Scene 3.

PUTTOCK (A Kite).

Imogen. . . . I chose an eagle,
And did avoid a puttock.
CYMBELINE, Act i. Scene 2.

OWL.

Ariel. There I couch when owls do cry.
THE TEMPEST, Act v. Scene 1.

King Richard. For night-owls shriek, where mounting larks should sing.
KING RICHARD II., Act. iii. Scene 3.

Warwick. Our soldiers'—like the night-owl's lazy flight,
Or like a lazy thresher with a flail—
Fell gently down, as if they struck their friends.

King Henry. The owl shriek'd at thy birth, an evil sign;
KING HENRY VI., Part III. Act ii. Scene 1;
Act v. Scene 6.

Ajax. I bade the vile owl go learn me the tenor of the proclamation, and he rails upon me.
TROILUS AND CRESSIDA, Act ii. Scene 1.

Ophelia. . . . They say, the owl was a baker's daughter.
HAMLET, Act iv. Scene 5.

Arviragus. The night to the owl, and morn to the lark, less welcome.
CYMBELINE, Act iii. Scene 6.

Lady Macbeth. It was the owl that shriek'd, the fatal bellman
Which gives the stern'st good night.
MACBETH, Act ii. Scene 2.

CHOUGH.

Puck. Or russet-pated choughs, many in sort,
Rising and cawing at the gun's report,
MIDSUMMER NIGHT'S DREAM, Act iii. Scene 2.

Edgar. . . . How fearful
And dizzy 'tis, to cast one's eyes so low!
The crows, and choughs, that wing the midway air,
Show scarce so gross as beetles;
KING LEAR, Act iv. Scene 6.

CROW.

Demetrius. That pure congealed white, high Taurus' snow,
Fann'd with the eastern wind, turns to a crow,
When thou hold'st up thy hand:
MIDSUMMER NIGHT'S DREAM, Act iii. Scene 2.

Paulina. The casting forth to crows thy baby daughter,
>> A Winter's Tale., Act iii. Scene 2.

Iden. . . . . cut off thy most ungracious head ;
Which I will bear in triumph to the king,
Leaving thy trunk for crows to feed upon.
>> King Henry VI., Part II. Act iv. Scene 10.

Troilus. . . . the busy day,
Waked by the lark, hath roused the ribald crows,
>> Troilus and Cressida, Act iv. Scene 2.

2d Captain. A leg of Rome shall not return to tell
What crows have peck'd them here :
>> Cymbeline, Act v. Scene iii.

Coriolanus. . . . —thus we debase
The nature of our seats, and make the rabble
Call our cares, fears : which will in time
Break ope the locks o' the senate, and bring in
The crows to peck the eagles.
>> Coriolanus, Act iii. Scene 1.

Macbeth. . . . Light thickens ; and the crow
Makes wing to the rooky wood ;
>> Macbeth, Act iii. Scene 2.

And thou, treble-dated crow,
That thy sable gender mak'st
With the breath thou giv'st and tak'st,
'Mongst our mourners shalt thou go.
>> Phœnix and the Turtle.

RAVEN.

Caliban. As wicked dew as e'er my mother brush'd
With raven's feather from unwholesome fen,
Drop on you both!
The Tempest, Act i. Scene 2.

Pistol. Young ravens must have food.
Merry Wives of Windsor, Act i. Scene 3.

Hotspur. . . . sometime he angers me,
With telling me of . . . .
. . . .
A clip-wing'd griffin, and a moulten raven,
King Henry IV., Part I. Act iii. Scene 1.

King Henry. What, doth my lord of Suffolk comfort me?
Came he right now to sing a raven's note,
Whose dismal tune bereft my vital powers;
King Henry VI., Part II. Act iii. Scene 2.

King Henry. The raven rook'd her on the chimney's top,
King Henry VI., Part III. Act v. Scene 6.

Thersites. . . . I would croak like a raven;
I would bode, I would bode.
Troilus and Cressida, Act v. Scene 2.

Juliet. For thou wilt lie upon the wings of night
Whiter than new snow on a raven's back.—
Romeo and Juliet, Act iii. Scene 2.

Iachimo. Swift, swift, you dragons of the night,
    that dawning
May bare the raven's eye!
                    CYMBELINE, Act ii. Scene 2.

Othello. . . . O, it comes o'er my memory,
As doth the raven o'er the infected house,
Boding to all,—
                    OTHELLO, Act iv. Scene 1.

Lady Macbeth. . . . The raven himself is
    hoarse
That croaks the fatal entrance of Duncan
Under my battlements.
                    MACBETH, Act i. Scene 5.

MAGPIE.

King Henry. And chattering pies in dismal dis-
    cords sung.
            KING HENRY VI., Part III. Act v. Scene 6.

ROOK.

Macbeth. . . . they say, blood will have
    blood;
Stones have been known to move, and trees to
    speak;
Augurs, and understood relations, have
By magot-pies, and choughs, and rooks, brought
    forth
The secret'st man of blood.
                    MACBETH, Act iii. Scene 4.

KITE.

Antigonus. . . . Come on, poor babe:
Some powerful spirit instruct the kites and ravens
To be thy nurses!
    A WINTER'S TALE, Act ii. Scene 3.

Hastings. More pity that the eagle should be
 mew'd,
While kites and buzzards prey at liberty.
    KING RICHARD III., Act i. Scene 1.

Antony. . . . —Ay, you kite!—
    ANTONY AND CLEOPATRA, Act iii. Scene 11.

3d Servant. Where dwellest thou?
Coriolanus. Under the canopy.
3d Servant. Under the canopy?
Coriolanus. Ay.
3d Servant. Where's that?
Coriolanus. I' the city of kites and crows.
    CORIOLANUS, Act iv. Scene 5.

Macbeth. . . . . . .
If charnel-houses, and our graves, must send
Those that we bury, back, our monuments
Shall be the maws of kites.
    MACBETH, Act iii. Scene 4.

DAW.

Warwick. But in these nice sharp quillets of the
 law,
Good faith, I am no wiser than a daw.
    KING HENRY VI., Part I. Act ii. Scene 4.

Pandarus. . . . Ne'er look, ne'er look; the eagles are gone; crows and daws, crows and daws! I had rather be such a man as Troilus, than Agamemnon and all Greece.
    Troilus and Cressida, Act i. Scene 2.

Iago. . . . I will wear my heart upon my sleeve
For daws to peck at.
    Othello, Act i. Scene 1.

3d Servant. . . . Then thou dwellest with daws too?
Coriolanus. No, I serve not thy master.
    Coriolanus, Act iv. Scene 5.

STARLING.

Hotspur. . . . I'll have a starling shall be taught to speak
Nothing but *Mortimer*, and give it him,
    King Henry IV., Part I. Act i. Scene 3.

JAY.

Caliban. I pr'ythee let me bring thee where crabs grow,
And I with my long nails will dig thee pig nuts;
Show thee a jay's nest,
    The Tempest, Act ii. Scene 2.

Mrs. Ford. We'll teach him to know turtles from jays.
    Merry Wives of Windsor, Act iii. Scene 3.

Petrucio. What, is the jay more precious than the lark,
Because his feathers are more beautiful?
Taming of the Shrew, Act iv. Scene 3.

PHEASANT.

Clown. Advocate's the court-word for a pheasant; say, you have none.

Shepherd. None, sir; I have no pheasant, cock nor hen.
A Winter's Tale, Act iv. Scene 3.

PARTRIDGE.

Beatrice. . . . he'll but break a comparison or two on me; which, peradventure, not marked, or not laughed at, strikes him into melancholy; and then there's a partridge wing saved, for the fool will eat no supper that night.
Much Ado about Nothing, Act ii. Scene 1.

Warwick. Who finds the partridge in the puttock's nest,
But may imagine how the bird was dead,
Although the kite soar with unblooded beak?
King Henry VI., Part II. Act iii. Scene 2.

QUAIL.

Thersites. Here's Agamemnon,—an honest fellow enough, and one that loves quails;
Troilus and Cressida, Act v. Scene 1.

WOODCOCK.

Claudio. Shall I not find a woodcock too?
        Much Ado about Nothing, Act v. Scene 1.

Fabian. Now is the woodcock near the gin.

Clown. . . . thou shalt hold the opinion of Pythagoras ere I will allow of thy wits; and fear to kill a woodcock, lest thou dispossess the soul of thy grandam.
        Twelfth Night, Act ii. Scene 5;
        Act iv. Scene 2.

Grumio. O this woodcock! what an ass it is!
        Taming of the Shrew, Act i. Scene 2.

1st Lord. Go, tell the count Rousillon, and my brother,
We have caught the woodcock, and will keep him muffled
Till we do hear from them.
        All's Well that Ends Well, Act iv. Scene 1.

Biron. . . . I have my wish;
Dumain transform'd: four woodcocks in a dish!
        Love's Labour's Lost, Act iv. Scene 3.

Clifford. . . . so strives the woodcock with the gin.
        King Henry VI., Part III. Act i. Scene 4.

Polonius. . . . springes to catch woodcocks.

Osric. How is't, Laertes?
Laertes. Why, as a woodcock to mine own springe, Osric;
I am justly kill'd with mine own treachery.
HAMLET, Act i. Scene 3; Act v. Scene 2.

PIGEON.

Celia. . . . .
Here comes monsieur le Beau.
Rosalind. With his mouth full of news.
Celia. Which he will put on us, as pigeons feed their young.
AS YOU LIKE IT, Act i. Scene 2.

Salarino. O, ten times faster Venus' pigeons fly
To seal love's bonds new made, than they are wont
To keep obligèd faith unforfeited.
MERCHANT OF VENICE, Act ii. Scene 6.

Shallow. . . . But for William cook;—are there no young pigeons?
KING HENRY IV., Part II. Act v. Scene 1.

DOVE.

Bottom. . . . I will aggravate my voice so, that I will roar you as gently as any sucking dove;
MIDSUMMER NIGHT'S DREAM, Act i. Scene 2.

Gobbo. I have here a dish of doves, that I would bestow upon your worship;
MERCHANT OF VENICE, Act ii. Scene 2.

Petrucio. For she's not forward, but modest as the dove;
>> TAMING OF THE SHREW, Act ii. Scene 1.

King Henry. Our kinsman Gloster is as innocent
From meaning treason to our royal person,
As is the sucking lamb, or harmless dove:

. . . .

Queen Margaret. Seems he a dove? his feathers are but borrowed,
For he's disposed as the hateful raven:
>> KING HENRY VI., Part II. Act iii. Scene 1.

Paris. He eats nothing but doves,
>> TROILUS AND CRESSIDA, Act iii. Scene 1.

Romeo. So shows a snowy dove trooping with crows,
As yonder lady o'er her fellows shows.
>> ROMEO AND JULIET, Act i. Scene 5.

Fair was the morn, when the fair queen of love,

. . . .

Paler for sorrow than her milk-white dove,
>> PASSIONATE PILGRIM, vii.

NIGHTINGALE.

Valentine. Except I be by Silvia in the night,
There is no music in the nightingale;

Valentine. Here can I sit alone, unseen of any,
And to the nightingale's complaining notes
Tune my distresses, and record my woes.
>> TWO GENTLEMEN OF VERONA, Act iii. Scene 1;
>> Act v. Scene 4.

Bottom. . . . I will roar you an't were any nightingale.
     MIDSUMMER NIGHT'S DREAM, Act i. Scene 2.

Portia. . . . I think,
The nightingale, if she should sing by day,
When every goose is cackling, would be thought
No better a musician than the wren.
     MERCHANT OF VENICE, Act v. Scene 1.

Lord. Wilt thou have music? hark! Apollo plays,
And twenty cagèd nightingales do sing:

Petrucio. Say, that she rail; why, then I'll tell her plain
She sings as sweetly as a nightingale:
    TAMING OF THE SHREW, Ind., Scene 2;
     Act ii. Scene 1.

Juliet. Wilt thou be gone? it is not yet near day:
It was the nightingale, and not the lark,
That pierced the fearful hollow of thine ear;
Nightly she sings on yon pomegranate-tree:
Believe me, love, it was the nightingale.
    ROMEO AND JULIET, Act iii. Scene 5.

Edgar. The foul fiend haunts poor Tom in the voice of a nightingale.
     KING LEAR, Act iii. Scene 6.

  As it fell upon a day,
  In the merry month of May,
  Sitting in a pleasant shade
  Which a grove of myrtles made,
  Beasts did leap, and birds did sing,
  Trees did grow, and plants did spring:

Everything did banish moan,
Save the nightingale alone:
She, poor bird, as all forlorn,
Lean'd her breast up-till a thorn,
And there sung the dolefull'st ditty,
That to hear it was great pity:
"Fie, fie, fie," now would she cry,
"Tereu, Tereu!" by-and-by:
That to hear her so complain,
Scarce I could from tears refrain;
For her griefs so lively shown,
Made me think upon mine own.
             PASSIONATE PILGRIM, xix.

LARK.

Helena. Your eyes are load-stars; and your tongue's sweet air
More tunable than lark to shepherd's ear,
             MIDSUMMER NIGHT'S DREAM, Act i. Scene 1.

When shepherds pipe on oaten straws,
And merry larks are ploughmen's clocks,
             LOVE'S LABOUR'S LOST, Act v. Scene 2.

Portia. The crow doth sing as sweetly as the lark,
When neither is attended;
             MERCHANT OF VENICE, Act v. Scene 1.

Autolycus. The lark that tirra-lirra chants,
             WINTER'S TALE, Act iv. Scene 2.

Dauphin. Nay, the man hath no wit that cannot, from the rising of the lark to the lodging of the lamb, vary deserved praise on my palfrey:
             KING HENRY V., Act iii. Scene 7.

Old Lady. With your theme, I could
O'ermount the lark.
>> KING HENRY VIII., Act ii. Scene 3.

Romeo. It was the lark, the herald of the morn,

. . . .

Juliet. It is, it is, hie hence, be gone, away;
It is the lark that sings so out of tune,
Straining harsh discords, and unpleasing sharps.
Some say the lark makes sweet division;
This doth not so, for she divideth us:
>> ROMEO AND JULIET, Act iii. Scene 5.

Edgar. From the dread summit of this chalky
bourn:
Look up a-height;—the shrill-gorged lark so far
Cannot be seen or heard:
>> KING LEAR, Act iv. Scene 6.

Hark! hark! the lark at heaven's gate sings,
>> CYMBELINE, Act ii. Scene 3.

Lo! here the gentle lark, weary of rest,
From his moist cabinet mounts up on high,
And wakes the morning, from whose silver breast
The sun ariseth in his majesty;
>> VENUS AND ADONIS.

Like to the lark at break of day arising
From sullen earth sings hymns at heaven's gate;
>> SONNET XXIX.

ROBIN.

Speed. . . . to relish a love-song, like a robin-redbreast;
>> TWO GENTLEMEN OF VERONA, Act ii. Scene 1.

Arviragus. With fairest flowers,
Whilst summer lasts, and I live here, Fidele,
I'll sweeten thy sad grave: thou shalt not lack
The flower that's like thy face, pale primrose; nor
The azured hare-bell, like thy veins; no, nor
The leaf of eglantine, whom not to slander,
Out-sweeten'd not thy breath: the ruddock would
With charitable bill, (O bill, sore-shaming
Those rich-left heirs that let their fathers lie
Without a monument!) bring thee all this;
Yea, and furr'd moss besides, when flowers are none,
To winter-ground thy corse.
CYMBELINE, Act iv. Scene 2.

THROSTLE. THRUSH.

Bottom. The throstle with his note so true,
MIDSUMMER NIGHT'S DREAM, Act iii. Scene 1.

Portia. . . . if a throstle sing he falls straight
a capering;
MERCHANT OF VENICE, Act i. Scene 2.

Autolycus. With heigh! with heigh! the thrush
and the jay:
Are summer songs for me and my aunts,
While we lie tumbling in the hay.
WINTER'S TALE, Act iv. Scene 2.

OUSEL-COCK.

Bottom. The ousel-cock, so black of hue,
With orange-tawny bill,
MIDSUMMER NIGHT'S DREAM, Act iii. Scene 1.

Silence. Alas! a black ouzel, cousin Shallow.
King Henry IV., Part II. Act iii. Scene 2.

CUCKOO.

Pistol. Take heed, ere summer comes, or cuckoo birds do sing.—
Merry Wives of Windsor, Act ii. Scene 1.

Bottom. . . . . . .
    The plain-song cuckoo gray,
    Whose note full many a man doth mark,
    And dares not answer, nay—
for indeed, who would set his wit to so foolish a bird? who would give a bird the lie, though he cry *cuckoo* never so?
Midsummer Night's Dream, Act iii. Scene 1.

Lorenzo.   That is the voice,
Or I am much deceived, of Portia.
    Portia. He knows me, as the blind man knows the cuckoo,
By the bad voice.
Merchant of Venice, Act v. Scene 1.

King Henry. So, when he had occasion to be seen,
He was but as the cuckoo is in June,
Heard, not regarded;
King Henry IV., Part I. Act iii. Scene 2.

Pompey. . . . the cuckoo builds not for himself,
Antony and Cleopatra Act ii. Scene 6.

LAPWING.

Adriana. Far from her nest the lapwing cries, away;
COMEDY OF ERRORS, Act iv. Scene 2.

Hero. For look where Beatrice, like a lapwing, runs
Close by the ground,
MUCH ADO ABOUT NOTHING, Act iii. Scene 1.

Horatio. This lapwing runs away with the shell on his head.
HAMLET, Act v. Scene 2.

FINCH.

Bottom. The finch, the sparrow, and the lark,
MIDSUMMER NIGHT'S DREAM, Act iii. Scene 1.

SWALLOW.

Perdita. . . . . daffodils,
That come before the swallow dares,
WINTER'S TALE, Act iv. Scene 3.

Falstaff. . . . Do you think me a swallow, an arrow or a bullet?
KING HENRY IV., Part II. Act iv. Scene 3.

Richmond. True hope is swift, and flies with swallow's wings,
Kings it makes gods, and meaner creatures kings.
KING RICHARD III., Act v. Scene 2.

Scarus. Swallows have built
In Cleopatra's sails their nests:
ANTONY AND CLEOPATRA, Act iv. Scene 10.

2d Lord. The swallow follows not summer
More willing than we your lordship.
TIMON OF ATHENS, Act iii. Scene 6.

Titus. And I have horse will follow where the game
Makes way, and run like swallows o'er the plain.
TITUS ANDRONICUS, Act ii. Scene 2.

MARTLET.

Banquo. This guest of summer,
The temple-haunting martlet, does approve,
By his loved mansionry, that the heaven's breath
Smells wooingly here: no jutty, frieze,
Buttress, nor coigne of vantage, but this bird
Hath made his pendent bed, and procreant cradle:
Where they most breed and haunt, I have observed,
The air is delicate.
MACBETH, Act i. Scene 6.

HEDGE SPARROW.

Fool. The hedge-sparrow fed the cuckoo so long,
That it had its head bit off by its young.
KING LEAR, Act i. Scene 4.

SPARROW.

Ceres. Tell me, heavenly bow,

If Venus, or her son, as thou dost know,
Do now attend the queen? . . . .

    Iris.   .   .   .   .   .   .
Her waspish-headed son has broke his arrows,
Swears he will shoot no more, but play with sparrows,
And be a boy right out.
THE TEMPEST, Act iv. Scene 1.

    Lucio. . . . sparrows must not build in his house-eaves,
MEASURE FOR MEASURE, Act iii. Scene 2.

    Adam. . . . He that doth the ravens feed,
Yea, providently caters for the sparrow,
Be comfort to my age!
AS YOU LIKE IT, Act ii. Scene 3.

    Prince Henry. He that rides at high speed, and with his pistol kills a sparrow flying.
    Falstaff. You have hit it.
    Prince Henry. So did he never the sparrow.
KING HENRY IV., Part I. Act ii. Scene 4.

    Hamlet. . . . there's a special providence in the fall of a sparrow.
HAMLET, Act v. Scene 2.

    Thersites. . . . I will buy nine sparrows for a penny, and his *pia mater* is not worth the ninth part of a sparrow.

    Pandarus. . . . —she fetches her breath as short as a new-ta'en sparrow.
TROILUS AND CRESSIDA, Act ii. Scene 1;
Act iii. Scene 2.

Duncan.   Dismay'd not this
Our captains, Macbeth and Banquo?
  Soldier.   Yes;
As sparrows, eagles;
                         MACBETH, Act i. Scene 2.

WREN.

Bottom. The wren with little quill;
       MIDSUMMER NIGHT'S DREAM, Act iii. Scene 1.

Sir Toby. Look where the youngest wren of nine comes.
                         TWELFTH NIGHT, Act iii. Scene 2.

King Henry. . . . thinks he that the chirping
     of a wren,
By crying comfort from a hollow breast,
Can chase away the first-conceivèd sound?
            KING HENRY VI., Part II. Act iii. Scene 2.

Gloucester. . . . the world is grown so bad
That wrens make prey where eagles dare not perch:
                KING RICHARD III., Act i. Scene 3.

Dionyza.   Be one of those that think
The pretty wrens of Tharsus will fly hence,
And open this to Pericles.
                         PERICLES, Act iv. Scene 4.

Lady Macduff. . . . the poor wren,
The most diminutive of birds, will fight,
Her young ones in her nest, against the owl.
                         MACBETH, Act iv. Scene 2.

SWAN. CYGNET.

Celia. . . . if she be a traitor,
Why so am I; we still have slept together,
Rose at an instant, learn'd, play'd, eat together;
And wheresoe'er we went, like Juno's swans,
Still we went coupled, and inseparable.
As You Like It, Act i. Scene 3.

Portia. . . . . . .
Let music sound, while he doth make his choice;
Then, if he lose, he makes a swan-like end,
Fading in music:
Merchant of Venice, Act iii. Scene 2.

Prince Henry. I am the cygnet to this pale faint swan,
Who chants a doleful hymn to his own death;
King John, Act v. Scene 7.

Suffolk. So doth the swan her downy cygnets save,
Keeping them prisoner underneath her wings.
King Henry VI., Part I. Act v. Scene 3.

Antony. . . . the swan's down feather
That stands upon the swell at full of tide,
And neither way inclines.
Antony and Cleopatra, Act iii. Scene 2.

Emilia. . . . I will play the swan,
And die in music;—
Othello, Act v. Scene 2.

Benvolio. Compare her face with some that I shall show,
And I will make thee think thy swan a crow.
> ROMEO AND JULIET, Act i. Scene 2.

Imogen. I' the world's volume
Our Britain seems as of it, but not in it;
In a great pool, a swan's nest.
> CYMBELINE, Act iii. Scene 4.

Aaron. For all the water in the ocean
Can never turn the swan's black legs to white,
Although she lave them hourly in the flood:
> TITUS ANDRONICUS, Act iv. Scene 2.

Let the priest in surplice white,
That defunctive music can,
Be the death-divining swan,
Lest the requiem lack his right.
> THE PHŒNIX AND THE TURTLE.

OSPREY.

Aufidius. I think he'll be to Rome,
As is the osprey to the fish, who takes it
By sovereignty of nature.
> CORIOLANUS, Act iv. Scene 7.

HANDSAW (A HERON).

Hamlet. . . . when the wind is southerly I know a hawk from a handsaw.
> HAMLET, Act ii. Scene 2.

PELICAN.

Gaunt. O, spare me not, my brother Edward's son,
For that I was his father Edward's son;
That blood already, like the pelican,
Hast thou tapp'd out,
    KING RICHARD II., Act ii. Scene 1.

Laertes. To his good friends thus wide I'll ope my
 arms;
And, like the kind life-rendering pelican,
Repast them with my blood.
    HAMLET, Act iv. Scene 5.

GULL.

Worcester. And being fed by us, you used us so
As that ungentle gull the cuckoo's bird
Useth the sparrow;
    KING HENRY IV., Part I. Act v. Scene 1.

Senator. . . . I do fear,
When every feather sticks in his own wing,
Lord Timon will be left a naked gull,
Which flashes now a phœnix.
    TIMON OF ATHENS, Act ii. Scene 1.

SCAMEL (SEA-MELL).

Caliban. . . . I'll bring thee
To clust'ring filberds, and sometimes I'll get thee
Young scamels from the rock.
    THE TEMPEST, Act ii. Scene 2.

MALLET. MALLARD.

Falstaff. . . . there is no more conceit in him than is in a mallet.
KING HENRY IV., Part II. Act ii. Scene 4.

Scarus. Claps on his sea-wing, and like a doting mallard,
ANTONY AND CLEOPATRA, Act iii. Scene 8.

WILD DUCK.

Falstaff. . . . there's no more valour in that Poins than in a wild duck.

Falstaff. . . . such as fear the report of a caliver worse than a struck fowl, or a hurt wild duck.
KING HENRY IV., Part I. Act ii. Scene 2;
Act iv. Scene 2.

DIVE-DAPPER.

Like a dive-dapper peering through a wave,
Who, being look'd on, ducks as quickly in;
VENUS AND ADONIS.

WILD-FOWL.

Clown. What is the opinion of Pythagoras concerning wild-fowl?
TWELFTH NIGHT, Act iv. Scene 2.

Bottom. . . . a lion among ladies, is a most dreadful thing: for there is not a more fearful wild-fowl than your lion, living;
   MIDSUMMER NIGHT'S DREAM, Act iii. Scene 1.

WILD GOOSE.

Puck. . . . when they him spy,
As wild geese that the creeping fowler eye,
   MIDSUMMER NIGHT'S DREAM, Act iii. Scene 2.

Jaques. . . . if he be free,
Why then, my taxing like a wild goose flies,
Unclaimed of any man.
   AS YOU LIKE IT, Act ii. Scene 7.

Falstaff. A king's son! If I do not beat thee out of thy kingdom with a dagger of lath, and drive all thy subjects afore thee like a flock of wild geese, I'll never wear hair on my face more.
   KING HENRY IV., Part I. Act ii. Scene 4.

GOOSE.

Launce. . . . I have stood on the pillory for geese he hath killed, otherwise he had suffered for't:
   TWO GENTLEMEN OF VERONA, Act iv. Scene 4.

Slender. . . . —Pray you, uncle, tell mistress Anne the jest, how my father stole two geese out of a pen, good uncle.
   MERRY WIVES OF WINDSOR, Act iii. Scene 4.

Biron. The spring is near, when green geese are a-breeding.
LOVE'S LABOUR'S LOST, Act i. Scene 1.

Mercutio. Nay, if our wits run the wild-goose chase, I am done; for thou hast more of the wild-goose in one of thy wits, than, I am sure, I have in my whole five. Was I with you there for the goose?
Romeo. Thou wast never with me for anything, when thou wast not there for the goose.
Mercutio. I will bite thee by the ear for that jest.
Romeo. Nay, good goose, bite not.
Mercutio. Thy wit is a very bitter sweeting; it is a most sharp sauce.
Romeo. And is it not well served in to a sweet goose?
Mercutio. O, here's a wit of cheverel that stretches from an inch narrow to an ell broad!
Romeo. I stretch it out for that word — broad: which added to the goose, proves thee far and wide a broad goose.
ROMEO AND JULIET, Act ii. Scene 4.

GOSLING.

Coriolanus. . . . I'll never
Be such a gosling to obey instinct;
CORIOLANUS, Act v. Scene 3.

DUCK.

Stephano. . . . swear then how thou escapedst.

Trinculo. Swam ashore, man, like a duck; I can swim like a duck, I'll be sworn.
Stephano. Here, kiss the book. Though thou canst swim like a duck, thou art made like a goose.
THE TEMPEST, Act ii. Scene 2.

Gower. . . . the grizzled north
Disgorges such a tempest forth,
That, as a duck for life that dives,
So up and down the poor ship drives.
PERICLES, Act 3.

TURKEY.

Fabian. . . . Contemplation makes a rare turkey-cock of him;
TWELFTH NIGHT, Act ii. Scene 5.

1st Carrier. The turkeys in my pannier are quite starved.
KING HENRY IV., Part I. Act ii. Scene 1.

Gower. Why, here he comes, swelling like a turkey-cock.
Fluellan. 'Tis no matter for his swellings, nor his turkey-cocks.—Got pless you, ancient Pistol!
KING HENRY V., Act v. Scene 1.

COCK.

Ariel. Hark, hark! I hear
　　The strain of strutting chanticleer
　　Cry, Cock-a-doodle-doo.
THE TEMPEST, Act i. Scene 2.

Speed. . . . You were wont, when you laughed,
to crow like a cock;
Two Gentlemen of Verona, Act ii. Scene 1.

Oberon. And look thou meet me ere the first cock crow.
Midsummer Night's Dream, Act ii. Scene 1.

Jacques. When I did hear
The motley fool thus moral on the time,
My lungs began to crow like chanticleer,
As You Like It, Act ii. Scene 7.

Katharine. What is your crest? A coxcomb?
Petrucio. A combless cock, so Kate will be my hen.
Katharine. No cock of mine, you crow too like a craven.
Taming of the Shrew, Act ii. Scene 1.

Horatio. . . . I have heard,
The cock, that is the trumpet to the morn,
Doth with his lofty and shrill-sounding throat
Awake the god of day; and, at his warning,
Whether in sea or fire, in earth or air,
The extravagant and erring spirit hies
To his confine:
Marcellus. . . . . .
Some say, that ever 'gainst that season comes
Wherein our Saviour's birth is celebrated,
The bird of dawning singeth all night long;
Hamlet, Act i. Scene 1.

Porter. 'Faith, sir, we were carousing till the second cock:
Macbeth, Act ii. Scene 3.

HEN.

Falstaff. How now, dame Partlet the hen?
KING HENRY IV., Part I. Act iii. Scene 3.

Shallow. . . . —Some pigeons, Davy; a couple of short-legged hens; a joint of mutton; and any pretty little tiny kickshaws, tell William cook.
KING HENRY IV., Part II. Act v. Scene 1.

Volumnia. . . . Thou hast never in thy life
Show'd thy dear mother any courtesy;
When she, (poor hen!) fond of no second brood,
Has cluck'd thee to the wars, and safely home,
Loaden with honour.
CORIOLANUS, Act v. Scene 3.

CHICKEN.

Prospero. . . My Ariel;—chick,—
THE TEMPEST, Act v. Scene 1.

York. . . . .
Say as you think, and speak it from your souls,—
Were't not all one, an empty eagle were set
To guard the chicken from a hungry kite,
KING HENRY VI., Part II. Act iii. Scene 1.

Pandarus. Troilus? why, he esteems her no more than I esteem an addle egg.
Cressida. If you love an addle egg as well as you love an idle head, you would eat chickens i' the shell.
TROILUS AND CRESSIDA, Act i. Scene 2.

All Servants. Gramercies, good fool: how does your mistress?

Fool. She's e'en setting on water to scald
Such chickens as you are.
TIMON OF ATHENS, Act ii. Scene 2.

Posthumus. . . . forthwith, they fly
Chickens, the way which they stoop'd eagles;
CYMBELINE, Act v. Scene 3.

Macduff. . . . . . .
What, all my pretty chickens,
MACBETH, Act iv. Scene 3.

FOWL.

Dromio E. . . . I pray thee, let me in.

Dromio S. (*within*). Ay, when fowls have no feathers, and fish have no fin.

Ant. E. Well, I'll break in : go, borrow me a crow.

Dromio E. A crow without feather; master, mean you so?
For a fish without a fin, there's a fowl without a feather :
If a crow help us in, sirrah, we'll pluck a crow together.
COMEDY OF ERRORS, Act iii. Scene 1.

Isabella. . . . Even for our kitchens
We kill the fowl of season :
MEASURE FOR MEASURE, Act ii. Scene 2.

Cardinal. Believe me, cousin Gloster,

Had not your man put up the fowl so suddenly,
We had had more sport.
>> KING HENRY VI., Part II., Act ii. Scene 1.

Gloster. Why, what a peevish fool was that of Crete,
That taught his son the office of a fowl;
And yet, for all his wings, the fool was drown'd!
>> KING HENRY VI., Part III. Act v. Scene 6.

Iachimo. . . . you know, strange fowl light upon neighbouring ponds.
>> CYMBELINE, Act i. Scene 5.

Marcus. . . . like a flight of fowl
Scatter'd by winds
>> TITUS ANDRONICUS, Act v. Scene 3.

ANIMALS.

Nathaniel. . . . he hath never fed of the dainties that are bred in a book; he hath not eat paper, as it were; he hath not drunk ink . . . he is only an animal only sensible in the duller parts;

LOVE'S LABOUR'S LOST, Act iv. Scene 2.

HORSE.

Puck. I jest to Oberon, and make him smile,
When I a fat and bean-fed horse beguile,
Neighing in likeness of a filly foal:

MIDSUMMER NIGHT'S DREAM, Act ii. Scene 1.

Benedict. I would my horse had the speed of your tongue; and so good a continuer:

MUCH ADO ABOUT NOTHING, Act i. Scene 1.

Orlando. . . . His horses are bred better; for, besides that they are fair with their feeding, they are taught their manage, and to that end riders dearly hired:

AS YOU LIKE IT, Act i. Scene 1.

Portia. Ay, that's a colt, indeed, for he doth nothing but talk of his horse;

MERCHANT OF VENICE, Act i. Scene 2.

Dion.   —Go,—fresh horses ;—
　　　　　A Winter's Tale, Act iii. Scene 1.

Curtis. . . . I pray thee, news?
Grumio. First, know, my horse is tired ; my master and mistress fallen out.
Curtis. How?
Grumio. Out of their saddles into the dirt. And thereby hangs a tale.
Curtis. Let's ha't, good Grumio.
Grumio. Lend thine ear.

. . . . . .

Now I begin : *Imprimis*, we came down a foul hill. my master riding behind my mistress :—
Curtis. Both of one horse?
Grumio. What's that to thee?
Curtis. Why, a horse.
Grumio. Tell thou the tale :—but hadst thou not crossed me, thou shouldst have heard how her horse fell, and she under her horse ; thou shouldst have heard, in how miry a place : how she was bemoiled ; how he left her with the horse upon her ; how he beat me because her horse stumbled ; how she waded through the dirt to pluck him off me ; how he swore ; how she prayed. that never prayed before ; how I cried ; how the horses ran away ; how her bridle was burst ; how I lost my crupper ; with many things of worthy memory, which now shall die in oblivion, and thou return unexperienced to thy grave.
　　　　　Taming of the Shrew, Act iv. Scene 1.

Ross. To horse, to horse ! urge doubts to them that fear.

Willoughby. Hold out my horse, and I will first be there.

. . . .

Groom. O, how it yearn'd my heart, when I beheld,
In London streets that coronation day,
When Bolingbroke rode on roan Barbary!
That horse that thou so often hast bestrid;
That horse that I so carefully have dress'd!
King Richard. Rode he on Barbary? Tell me, gentle friend,
How went he under him?
Groom. So proudly as if he had disdain'd the ground.

KING RICHARD II., Act ii. Scene 1;
Act v. Scene 5.

Hotspur. Hath Butler brought those horses from the sheriff?
Servant. One horse, my lord, he brought even now.
Hotspur. What horse? a roan, a crop-ear, is it not?
Servant. It is, my lord.
Hotspur. That roan shall be my throne.
Well, I will back him straight:
  Esperance!—
Bid Butler lead him forth into the park.

KING HENRY IV., Part I. Act ii. Scene 3.

North. Now, Travers, what good tidings come with you?
Travers. My lord, sir John Umfrevile turn'd me back
With joyful tidings; and, being better horsed,
Out-rode me. After him came, spurring hard,
A gentleman almost forspent with speed,

That stopp'd by me to breathe his bloodied horse :
He ask'd the way to Chester; and of him
I did demand what news from Shrewsbury.
He told me, that rebellion had ill luck,
And that young Harry Percy's spur was cold :
With that, he gave his able horse the head,
And, bending forward, struck his armed heels
Against the panting sides of his poor jade
Up to the rowel-head ; and starting so,
He seem'd in running to devour the way,
Staying no longer question.

  .   .   .   .

 North. Why should the gentleman that rode by
   Travers
Give then such instances of loss?
 Lord Bardolph.   Who, he?
He was some hilding fellow, that had stolen
The horse he rode on ;
    KING HENRY IV., Part II. Act i. Scene 1.

 Dauphin. . . . . he is, indeed,
a horse ; and all other jades you may call beasts.
    KING HENRY V., Act iii. Scene 7.

 Warwick. Between two horses, which doth bear
  him best,
    KING HENRY VI., Part I. Act ii. Scene 4.

 York.   .   .   .   .   .   .
. . as fast as horse can carry them ;
    KING HENRY VI., Part II. Act i. Scene 4.

 York.   .   .   .   .
Unless the adage must be verified,
That beggars, mounted, run their horse to death.
    KING HENRY VI., Part III. Act i. Scene 4.

Gloucester. But yet I run before my horse to market:

King Richard. A horse! a horse! my kingdom for a horse!
KING RICHARD III., Act i. Scene 1;
Act v. Scene 4.

Norfolk. . . . anger is like
A full-hot horse; who being allow'd his way,
Self-mettle tires him.
KING HENRY VIII., Act i. Scene 1.

Calphurnia. The noise of battle hurtled in the air,
Horses did neigh, and dying men did groan;
JULIUS CÆSAR, Act ii. Scene 2.

Antony. That which is now a horse, even with a thought
The rack dislimns, and makes it indistinct,
As water is in water.
ANTONY AND CLEOPATRA, Act iv. Scene 12.

Thersites. . . . I think thy horse will sooner con an oration, than thou learn a prayer without book.
TROILUS AND CRESSIDA, Act ii. Scene 1.

Marcius. Yonder comes news:—a wager they have met.
Lartius. My horse to yours, no.
Marcius.   'Tis done.
Lartius.   Agreed.
Marcius. Say, has our general met the enemy?
Messenger. They lie in view; but have not spoke as yet.

Lartius. So, the good horse is mine.
Marcius.   I'll buy him of you.
Lartius. No, I'll nor sell nor give him : lend you him I will,
For half a hundred years.
CORIOLANUS, Act i. Scene 4.

2d Servant. May it please your honour, the lord Lucius,
Out of his free love, hath presented to you
Four milk-white horses, trapp'd in silver.
TIMON OF ATHENS, Act i. Scene 2.

Hamlet. . . . six Barbary horses against six French swords,
HAMLET, Act v. Scene 2.

Mercutio.   This is that very Mab
That plats the manes of horses in the night;
ROMEO AND JULIET, Act i. Scene 4.

Fool. May not an ass know when the cart draws the horse?

Lear. It were a delicate stratagem, to shoe a troop of horse with felt:
KING LEAR, Act i. Scene 4; Act iv. Scene 6.

Macbeth. I wish your horses swift and sure of foot;
And so I do commend you to their backs.
Farewell.
MACBETH, Act iii. Scene 1.

DOG.

Ariel.   Hark, hark!
Bowgh-wowgh.
The watch-dogs bark:
Bowgh-wowgh.

TEMPEST, Act i. Scene 2.

Launce.   I think Crab my dog be the sourest-natured dog that lives: my mother weeping, my father wailing, my sister crying, our maid howling, our cat wringing her hands, and all our house in a great perplexity, yet did not this cruel-hearted cur shed one tear: he is a stone, a very pebble-stone, and has no more pity in him than a dog:

Launce.   She hath more qualities than a water-spaniel—which is much in a bare Christian.

Launce. When a man's servant shall play the cur with him, look you, it goes hard: one that I brought up of a puppy; one that I saved from drowning, when three or four of his blind brothers and sisters went to it!

Proteus. Where have you been these two days loitering?
Launce. Marry, sir, I carried mistress Silvia the dog you bade me.
Proteus. And what says she to my little jewel?
Launce. Marry, she says, your dog was a cur; and tells you, currish thanks is good enough for such a present.

Proteus. But she received my dog?

Launce. No, indeed did she not: here have I brought him back again.

Proteus. What, didst thou offer her this from me?

Launce. Ay, sir; the other squirrel was stolen from me by the hangman boys in the market-place: and then I offered her mine own; who is a dog as big as ten of yours, and therefore the gift the greater.

>   TWO GENTLEMEN OF VERONA, Act ii. Scene 3;
>   Act iii. Scene 1; Act iv. Scene 4.

Slender. How does your fallow greyhound, sir?

Slender. Why do your dogs bark so? be there bears i' the town?

>   MERRY WIVES OF WINDSOR, Act i. Scene 1.

Theseus. Go one of you, find out the forester;
For now our observation is perform'd;
And since we have the vaward of the day,
My love shall hear the music of my hounds.
Uncouple in the western valley; let them go:
Dispatch, I say, and find the forester.
We will, fair queen, up to the mountain's top,
And mark the musical confusion
Of hounds and echo in conjunction.

Hippolyta. I was with Hercules and Cadmus once,
When in a wood of Crete they bay'd the bear
With hounds of Sparta: never did I hear
Such gallant chiding; for, besides the groves,
The skies, the fountains, every region near
Seem'd all one mutual cry: I never heard
So musical a discord, such sweet thunder.

Theseus. My hounds are bred out of the Spartan kind

So flew'd, so sanded; and their heads are hung
With ears that sweep away the morning dew;
Crook-knee'd and dew-lapp'd like Thessalian bulls;
Slow in pursuit, but match'd in mouth like bells,
Each under each. A cry more tuneable
Was never holla'd to, nor cheer'd with horn,
In Crete, in Sparta, nor in Thessaly:
Judge, when you hear.
    MIDSUMMER NIGHT'S DREAM, Act iv. Scene 1.

Beatrice. . . . I had rather hear my dog bark at a crow, than a man swear he loves me

Benedict (aside). An he had been a dog that should have howled thus, they would have hanged him:
    MUCH ADO ABOUT NOTHING, Act i. Scene 1;
      Act ii. Scene 3.

Clown. Do not desire to see this letter.
Fabian. This is, to give a dog, and in recompense desire my dog again.
    TWELFTH NIGHT, Act v. Scene 1.

Celia. Why, cousin; why, Rosalind;—Cupid have mercy!—not a word?
Rosalind. Not one to throw at a dog.
    AS YOU LIKE IT, Act i. Scene 3.

Gratiano. . . . *I am sir Oracle,*
And when I ope my lips let no dog bark!
    MERCHANT OF VENICE, Act i. Scene 1.

Lord. Huntsman, I charge thee, tender well my hounds:
Brach Merriman,—the poor cur is emboss'd;

And couple Clowder with the deep-mouth'd brach.
Saw'st thou not, boy, how Silver made it good
At the hedge corner, in the coldest fault?
I would not lose the dog for twenty pound.

   1st Huntsman. Why, Belman is as good as he, my lord;
He cried upon it at the merest loss,
And twice to-day pick'd out the dullest scent.
Trust me, I take him for the better dog.

   Lord. Thou art a fool; if Echo were as fleet,
I would esteem him worth a dozen such.
But sup them well, and look unto them all;
To-morrow I intend to hunt again.
>   TAMING OF THE SHREW, Ind., Scene 1.

   Arthur. And, like a dog that is compelled to fight,
Snatch at his master that doth tarre him on.
>   KING JOHN, Act iv. Scene 1.

   Falstaff. . . . you may stroke him as gently as a puppy greyhound:
>   KING HENRY IV., Part II. Act ii. Scene 4.

   Ramburcs. That island of England breeds very valiant creatures; their mastiffs are of unmatchable courage.
>   KING HENRY V., Act iii. Scene 7.

   Talbot. They call'd us, for our fierceness, English dogs;

   Warwick. Between two dogs, which hath the deeper mouth,
>   KING HENRY VI., Part I. Act i. Scene 5;
>   Act ii. Scene 4.

Bolingbroke. The time when screech-owls cry, and ban-dogs howl,

Gloster. A staff is quickly found to beat a dog.
KING HENRY VI., Part II. Act i. Scene 4;
Act iii. Scene 1.

Richard. . . . . .
Or as a bear, encompass'd round with dogs;
Who having pinch'd a few, and made them cry
The rest stand all aloof, and bark at him.
KING HENRY VI., Part III. Act ii. Scene 1.

King Henry. To me you cannot reach, you play the spaniel,
And think with wagging of your tongue to win me;
KING HENRY VIII., Act v. Scene 3.

Brutus. And, gentle friends,
Let's kill him boldly, but not wrathfully;
Let's carve him as a dish fit for the gods,
Not hew him as a carcase fit for hounds:

Brutus. I had rather be a dog, and bay the moon,
Than such a Roman.
JULIUS CÆSAR, Act ii. Scene 1; Act iv. Scene iii.

Nestor. Two curs shall tame each other: pride alone
Must tarre the mastiffs on, as 'twere their bone.

Thersites. . . . he will spend his mouth, and promise, like Brabler the hound;
TROILUS AND CRESSIDA, Act i. Scene 3;
Act v. Scene 1.

Othello. Thou hadst better have been born a dog
Than answer my waked wrath.
                    OTHELLO, Act iii. Scene 3.

Marcius. They said they were an-hungry; sigh'd
    forth proverbs,—
That hunger broke stone walls; that dogs must eat;
                    CORIOLANUS, Act i. Scene 1.

3d Servant. . . . has sent your honour two
brace of greyhounds.

Timon. I am *Misanthropos*, and hate mankind.
For thy part, I do wish thou wert a dog,
That I might love thee something.
                    TIMON OF ATHENS, Act i. Scene 2;
                        Act iv. Scene 3.

Fool. Truth's a dog must to kennel;

Lear. The little dogs and all,
Tray, Blanch, and Sweetheart, see, they bark at me.
    Edgar. Tom will throw his head at them :—
            Avaunt, you curs!
Be thy mouth or black or white,
Tooth that poisons if it bite;
Mastiff, greyhound, mongrel grim,
Hound or spaniel, brach or lym;
Or bobtail tike, or trundle-tail;
Tom will make him weep and wail:
For, with throwing thus my head,
Dogs leap the hatch, and all are fled.

Lear. . . . Thou hast seen a farmer's dog
    bark at a beggar?

Gloster. Ay, sir.

Lear. And the creature run from the cur? There thou might'st behold the great image of authority: a dog's obey'd in office.

Cornelia. . . . Mine enemy's dog,
Though he had bit me, should have stood that night
Against my fire.

KING LEAR, Act i. Scene 4 ; Act iii. Scene 6 ;
Act iv. Scenes 6 and 7.

Cornelius. . . . She doth think she has
Strange lingering poisons : . . .
. . . Those she has
Will stupefy and dull the sense awhile :
Which first, perchance, she'll prove on cats and
    dogs ;

Cornelius. The queen, sir, very oft impórtuned
    me
To temper poisons for her ; still pretending
The satisfaction of her knowledge only
In killing creatures vile, as cats and dogs
Of no esteem :

CYMBELINE, Act i. Scene 5 ; Act v. Scene 5.

1st Murderer. We are men, my liege.
Macbeth. Ay, in the catalogue ye go for men ;
As hounds, and greyhounds, mongrels, spaniels, curs,
Shoughs, water-rugs, and demi-wolves, are cleped
All by the name of dogs :

MACBETH, Act iii. Scene 1.

ASS.

Adriana. There's none but asses will be bridled so.
                COMEDY OF ERRORS, Act ii. Scene 1.

Bottom. . . . I must to the barber's, monsieur; for, methinks, I am marvellous hairy about the face; and I am such a tender ass, if my hair do but tickle me I must scratch.

. . . .

Titania. My Oberon! what visions have I seen! Methought I was enamour'd of an ass.
                MIDSUMMER NIGHT'S DREAM, Act iv. Scene 1.

Conrade. Away! you are an ass, you are an ass.
Dogberry. Dost thou not suspect my place? Dost thou not suspect my years?—O that he were here to write me down, *an ass!* but, masters, remember, that I am *an ass;* though it be not written down, yet forget not that I am *an ass.*— . . . O, that I had been writ down—*an ass!*
                MUCH ADO ABOUT NOTHING, Act iv. Scene 2.

Katharine. Asses were made to bear, and so are you.
                TAMING OF THE SHREW, Act ii. Scene 1.

Bastard. It lies as sightly on the back of him,
As great Alcides shows upon an ass:—
But, ass, I'll take that burden from your back;
Or lay on that shall make your shoulders crack.
                KING JOHN, Act ii. Scene 1.

King Richard.   I was not made a horse;
And yet I bear a burden like an ass,
    KING RICHARD II., Act v. Scene 5.

Antony. And though we lay these honours on this man,
To ease ourselves of divers slanderous loads,
He shall but bear them as the ass bears gold,
    JULIUS CÆSAR, Act iv. Scene 1.

Iago. The Moor is of a free and open nature,
That thinks men honest that but seem to be so;
And will as tenderly be led by the nose, as asses are.
    OTHELLO, Act i. Scene 3.

Menemius. . . . your beards deserve not so honourable a grave as to stuff a botcher's cushion, or to be entombed in an ass's pack saddle.
    CORIOLANUS, Act ii. Scene 1.

2d Lord. Away, unpeaceable dog, or I'll spurn thee hence.
Apemantus. I will fly, like a dog, the heels of the ass.

Timon. . . . if thou wert the fox, the lion would suspect thee, when, peradventure, thou wert accused by the ass: if thou wert the ass, thy dulness would torment thee;
    TIMON OF ATHENS, Act i. Scene 1;
    Act iv. Scene 3.

Polonius. The actors are come hither, my lord.
Hamlet. Buz, buz!
Polonius. Upon mine honour,—
Hamlet. Then came each actor on his ass.—
    HAMLET, Act ii. Scene 2.

Cloten. . . . 'Would there had been some hurt done!

2d Lord (*aside*). I wish not so; unless it had been the fall of an ass, which is no great hurt.

                    CYMBELINE, Act i. Scene 2.

MULE.

Parolles. . . . Tongue, I must put you into a butter-woman's mouth, and buy myself another of Bajazet's mule, if you prattle me into these perils.

        ALL'S WELL THAT ENDS WELL, Act iv. Scene 1.

Alençon. Either they must be dieted like mules,
And have their provender tied to their mouths,
Or piteous they will look,

        KING HENRY VI., Part I. Act i. Scene 2.

Griffith. He fell sick suddenly, and grew so ill, He could not sit his mule.

            KING HENRY VIII., Act iv. Scene 2.

Soldier. . . . the messenger
Came on my guard; and at thy tent is now
Unloading of his mules.

        ANTONY AND CLEOPATRA, Act iv. Scene 6.

Brutus. . . . For an end,
We must suggest the people in what hatred
He still hath held them; that, to his power, he would
Have made them mules,

            CORIOLANUS, Act ii. Scene 1.

K

BULL, BULLOCK.

Sebastian. Whiles we stood here securing your repose,
Even now, we heard a hollow burst of bellowing
Like bulls,

Gonzalo. Dew-lapp'd like bulls,
TEMPEST, Act ii. Scene 1; Act iii. Scene 3.

Don Pedro. "In time the savage bull doth bear the yoke."
Benedict. The savage bull may; but if ever the sensible Benedick bear it, pluck off the bull's horns and set them in my forehead:
MUCH ADO ABOUT NOTHING, Act i. Scene 1.

Prince Henry. From a god to a bull? a heavy declension! it was Jove's case.
KING HENRY IV., Part II. Act ii. Scene 2.

Claudio. I wish him joy of her.
Benedict. Why, that's spoken like an honest drover; so they sell bullocks.
MUCH ADO ABOUT NOTHING, Act ii. Scene 1.

Shallow. . . . How a good yoke of bullocks at Stamford fair?
Silence. Truly, cousin, I was not there.
KING HENRY IV., Part II. Act iii. Scene 2.

OX.

Falstaff. I do begin to perceive that I am made an ass.
Ford. Ay, and an ox too;
MERRY WIVES OF WINDSOR, Act v. Scene 5.

Touchstone. As the ox hath his bow, sir, the horse his curb, and the falcon her bells, so man hath his desires;
AS YOU LIKE IT, Act iii. Scene 3.

Suffolk. So worthless peasants bargain for their wives,
As market-men for oxen, sheep, or horse.
KING HENRY VI., Part I. Act v. Scene 5.

George. Then is sin struck down like an ox, and iniquity's throat cut like a calf.
KING HENRY VI., Part II. Act iv. Scene 2.

Thersites. There's Ulysses and old Nestor,—whose wit was mouldy ere your grandsires had nails on their toes,—yoke you like draught oxen, and make you plough up the war.
TROILUS AND CRESSIDA, Act ii. Scene 1.

COW.

Beatrice. . . . it is said, *God sends a curst cow short horns:* but to a cow too curst he sends none.
MUCH ADO ABOUT NOTHING, Act ii. Scene 1.

Touchstone. . . . I remember the kissing of her batlet, and the cow's dugs that her pretty chopped hands had milked :
As You Like It, Act ii. Scene 4.

Scarus. The brize upon her, like a cow in June,—
Antony and Cleopatra, Act iii. Scene 8.

HEIFER.

Leontes. . . . Come, captain,
We must be neat ; not neat, but cleanly, captain :
And yet the steer, the heifer, and the calf
Are all call'd neat.—
A Winter's Tale, Act i. Scene 2.

Warwick. Who finds the heifer dead, and bleeding fresh,
And sees fast by a butcher with an axe,
But will suspect 'twas he that made the slaughter?
King Henry VI., Part II. Act iii. Scene 2.

CALF.

Holofernes. . . . he clepeth a calf, cauf ; half, hauf ;
Love's Labour's Lost, Act v. Scene 1.

Austria. King Philip, listen to the cardinal.
Bastard. And hang a calf's-skin on his recreant limbs.

Austria. Do so, king Philip; hang no more in doubt.
Bastard. Hang nothing but a calf's-skin, most sweet lout.
KING JOHN, Act iii. Scene 1.

King Henry. And as the butcher takes away the calf,
And binds the wretch, and beats it when it strays,
KING HENRY VI., Part II. Act iii. Scene 1.

Hamlet. And what did you enact?
Polonius. I did enact Julius Cæsar: I was killed i' the Capitol: Brutus killed me.
Hamlet. It was a brute part of him, to kill so capital a calf there.—
HAMLET, Act iii. Scene 2.

RAM.

Florizel. . . . Jupiter
Became a bull, and bellow'd; the green Neptune
A ram, and bleated;
A WINTER'S TALE, Act iv. Scene 3.

SHEEP.

Iris. Thy turfy mountains, where live nibbling sheep,
TEMPEST, Act iv. Scene 1.

Speed. Sir Proteus, save you! Saw you my master?
Proteus. But now he parted hence, to embark for Milan.

Speed. Twenty to one then he's shipp'd already; and I have play'd the sheep in losing him.

Proteus. Indeed a sheep doth very often stray, an if the shepherd be a while away.

Speed. You conclude that my master is a shepherd then, and I a sheep?

Proteus. I do.

Speed. Why then my horns are his horns, whether I wake or sleep.

Proteus. A silly answer, and fitting well a sheep.

Speed. This proves me still a sheep.

Proteus. True; and thy master a shepherd.

Speed. Nay, that I can deny by a circumstance.

Proteus. It shall go hard but I'll prove it by another.

Speed. The shepherd seeks the sheep, and not the sheep the shepherd; but I seek my master, and my master seeks not me: therefore, I am no sheep.

Proteus. The sheep for fodder follow the shepherd, the shepherd for food follows not the sheep; thou or wages followest thy master, thy master for wages follows not thee; therefore thou art a sheep.

Speed. Such another proof will make me cry baa.
  Two Gentlemen of Verona, Act i. Scene 1.

Benedict. Now, *Divine air!* now is his soul ravished!—Is it not strange that sheep's guts should hale souls out of men's bodies?—Well, a horn for my money, when all's done.
  Much Ado about Nothing, Act ii. Scene 3.

Armado (to Holofernes). Monsieur, are you not lettered?

Moth. Yes, yes; he teaches boys the horn-book;— What is a, b, spelt backward, with a horn on his head?

Holofernes. Ba, *pueritia*, with a horn added
Moth. Ba, most silly sheep, with a horn.—
        LOVE'S LABOUR'S LOST, Act v. Scene 1.

Shepherd. . . . Would any but these boiled brains of nineteen and two-and-twenty hunt this weather? They have scared away two of my best sheep; which, I fear, the wolf will sooner find than the master; if anywhere I have them, 'tis by the seaside, browsing of ivy.
        A WINTER'S TALE, Act iii. Scene 3.

Shallow. —How a score of ewes now?
Silence. Thereafter as they be: a score of good ewes may be worth ten pounds.
        KING HENRY IV., Part II. Act iii. Scene 2.

Talbot. Sheep run not half so treacherous from the wolf,
  .   .   .   .
As you fly from your oft-subdued slaves.
        KING HENRY VI., Part I. Act i. Scene 5.

Cade. They fell before thee like sheep.
        KING HENRY VI., Part II. Act iv. Scene 3.

Thersites. . . . I had rather be a tick in a sheep, than such a valiant ignorance.
        TROILUS AND CRESSIDA, Act iii. Scene 3.

LAMB.

Julia. Alas, poor Proteus! thou hast entertain'd
A fox, to be the shepherd of thy lambs:
        TWO GENTLEMEN OF VERONA, Act iv. Scene 4.

Messenger. . . . he hath borne himself beyond the promise of his age; doing, in the figure of a lamb, the feats of a lion :
>> MUCH ADO ABOUT NOTHING, Act i. Scene 1.

Duke. . . . my thoughts are ripe in mischief :
I'll sacrifice the lamb that I do love,
To spite a raven's heart within a dove.
>> TWELFTH NIGHT, Act v. Scene 1.

Gremio. . . . she's a lamb, a dove,
>> TAMING OF THE SHREW, Act iii. Scene 2.

Arthur. Nay, hear me, Hubert! drive these men away,
And I will sit as quiet as a lamb ;
>> KING JOHN, Act iv. Scene 1.

York. In peace, was never gentle lamb more mild.
>> KING RICHARD II., Act. ii. Scene 1.

King Henry. Our kinsman Gloster is as innocent
From meaning treason to our royal person,
As is the sucking lamb, or harmless dove :
Queen Margaret. . . . .
Is he a lamb? his skin is surely lent him.
>> KING HENRY VI., Part II. Act iii. Scene 1.

King Henry. And when the lion fawns upon the lamb,
The lamb will never cease to follow him.
>> KING HENRY VI., Part III. Act iv. Scene 8.

GOAT.

Falstaff. . . . Am I ridden with a Welsh goat too?
          Merry Wives of Windsor, Act v. Scene 5.

Touchstone. Come apace, good Audrey; I will fetch up your goats, Audrey :
          As You Like It, Act iii. Scene 3.

Shylock. A pound of man's flesh, taken from a man,
Is not so estimable, profitable neither,
As flesh of muttons, beefs, or goats.
          Merchant of Venice, Act i. Scene 3.

Glendower. The goats ran from the mountains, and the herds
Were strangely clamorous to the frighted fields.

Vernon. Wanton as youthful goats, wild as young bulls.
          King Henry IV., Part I. Act iii. Scene 1 ;
          Act iv. Scene 1.

Fluellen. . . . this leek ; . . .
. . I would desire you to eat it.
Pistol. Not for Cadwallader and all his goats.
Fluellen. There is one goat for you. [*Strikes him.*]
          King Henry V., Act v. Scene 1.

Arviragus. . . . what thing is it, that I never
Did see man die? scarce ever look'd on blood,
But that of coward hares, hot goats, and venison?
          Cymbeline, Act iv. Scene 4.

Aaron. I'll make you feed on berries, and on roots,
And feed on curds and whey, and suck the goat,
And cabin in a cave, and bring you up
To be a warrior, and command a camp.
TITUS ANDRONICUS, Act iv. Scene 2.

PIG.

Dromio E. The capon burns, the pig falls from the spit;
COMEDY OF ERRORS, Act i. Scene 2.

Shylock. Some men there are love not a gaping pig;
MERCHANT OF VENICE, Act iv. Scene 1.

Mercutio. O, then, I see, queen Mab has been with you.

. . . .

Sometimes she gallops o'er a courtier's nose,
And then dreams he of smelling out a suit:
And sometime comes she with a tithe-pig's tail,
Tickling a parson's nose as 'a lies asleep,
Then dreams he of another benefice:
ROMEO AND JULIET, Act i. Scene 4.

Aaron. Weke, weke! so cries a pig prepared to the spit.
TITUS ANDRONICUS, Act iv. Scene 2.

HOG.

Launcelot. . . . this making of Christians will raise the price of hogs; if we grow all to be pork-

eaters we shall not shortly have a rasher on the coals for money.
> MERCHANT OF VENICE, Act iii. Scene 5.

Edgar. . . . hog in sloth,
> KING LEAR, Act iii. Scene 4.

SOW.

Dauphin. I tell thee, constable, my mistress wears his own hair.
Constable. I could make as true a boast as that, if I had a sow to my mistress.
> KING HENRY V., Act iii. Scene 7.

BOAR.

Petrucio. Have I not heard the sea, puff'd up with winds,
Rage like an angry boar, chafèd with sweat?
> TAMING OF THE SHREW, Act i. Scene 2.

Hastings. To fly the boar, before the boar pursues,
Were to incense the boar to follow us,
> KING RICHARD III., Act iii. Scene 2.

. . . to-morrow he intends
To hunt the boar with certain of his friends.
> VENUS AND ADONIS.

STAG.

Falstaff. For me, I am here a Windsor stag;
> MERRY WIVES OF WINDSOR, Act v. Scene 5.

1st Lord. To the which place a poor sequester'd stag,
That from the hunters' aim had ta'en a hurt,
Did come to languish;
>As You Like It, Act ii. Scene 1.

Talbot. . . . moody-mad and desperate stags,
Turn on the bloody hounds with heads of steel,
And make the cowards stand aloof at bay:
>King Henry VI., Part I. Act iv. Scene 2.

Cæsar. Yea, like the stag, when snow the pasture sheets,
The barks of trees thou browsed'st;
>Antony and Cleopatra, Act i. Scene 4.

Lavinia. Jove shield your husband from his hounds to-day;
'Tis pity they should take him for a stag.
>Titus Andronicus, Act ii. Scene 3.

HIND.

Touchstone. If a hart do lack a hind,
Let him seek out Rosalind.
>As You Like It, Act iii. Scene 2.

DEER.

Mrs. Ford. . . . art thou there, my deer?
. . . .
. . I will never take you for my love again,
but I will always count you my deer.
>Merry Wives of Windsor, Act v. Scene 5.

Holofernes. Sir Nathaniel, will you hear an extemporal epitaph on the death of the deer? and, to humour the ignorant, I have called the deer the princess killed, a pricket.
>> Love's Labour's Lost, Act iv. Scene 2.

Duke S. And did you leave him in this contemplation?
2d Lord. We did, my lord, weeping and commenting upon the sobbing deer.
>> As You Like It, Act ii. Scene 1.

Talbot. A little herd of England's timorous deer,
>> King Henry VI., Part I. Act iv. Scene 2.

1st Keeper. Under this thick-grown brake we'll shroud ourselves;
For through this laund anon the deer will come;
And in this covert will we make our stand,
Culling the principal of all the deer.
>> King Henry VI., Part III. Act iii. Scene 1.

Antony. How like a deer, stricken by many princes,
Dost thou here lie!
>> Julius Cæsar, Act iii. Scene 1.

Cloten. "Tis gold
Which buys admittance; oft it doth; yea, and makes
Diana's rangers false themselves, yield up
Their deer to the stand o' the stealer;
>> Cymbeline, Act ii. Scene 3.

Rosse. Your castle is surprised; your wife, and babes,
Savagely slaughter'd: to relate the manner,
Were, on the quarry of these murder'd deer,
To add the death of you.
MACBETH, Act iv. Scene 3.

BUCK.

Pandarus. Love, love, nothing but love, still more!
For, oh, love's bow
Shoots buck and doe:
The shaft confounds,
Not that it wounds,
But tickles still the sore.
TROILUS AND CRESSIDA, Act iii. Scene 1.

DOE.

Demetrius. What, hast not thou full often struck a doe,
And borne her cleanly by the keeper's nose?

Chiron, we hunt not, we, with horse nor hound;
But hope to pluck a dainty doe to ground.
TITUS ANDRONICUS, Act ii. Scenes 1 and 2.

FAWN.

Orlando. Then, but forbear your food a little while,
Whiles, like a doe, I go to find my fawn,
And give it food.
AS YOU LIKE IT, Act ii. Scene 7.

HART.

Posthu. The swiftest harts have posted you by land :
CYMBELINE, Act ii. Scene 4.

ROE.

Princess. Whip to our tents, as roes run over land.
LOVE'S LABOUR'S LOST, Act v. Scene 2.

1st Servant. Say, thou wilt course ; thy greyhounds are as swift
As breathèd stags, ay, fleeter than the roe.
TAMING OF THE SHREW, Ind., Scene 2.

FOX.

Ford. . . . Here, here, here be my keys : ascend my chambers, search, seek, find out : I'll warrant we'll unkennel the fox.—
MERRY WIVES OF WINDSOR, Act iii. Scene 3.

Pompey. 'Twas never merry world since, of two usuries, the merriest was put down, and the worser allowed by order of law a furred gown to keep him warm ; and furred with fox and lamb-skins too, to signify, that craft, being richer than innocency, stands for the facing.
MEASURE FOR MEASURE, Act iii. Scene 2.

Lysander. This lion is a very fox for his valour.
Theseus. True ; and a goose for his discretion.

Demetrius. Not so, my lord; for his valour cannot carry his discretion; and the fox carries the goose.

Theseus. His discretion, I am sure, cannot carry his valour; for the goose carries not the fox.
MIDSUMMER NIGHT'S DREAM, Act v. Scene 1.

Moth. The fox, the ape, and the humble bee,
Were still at odds, being but three:
LOVE'S LABOUR'S LOST, Act iii. Scene 1.

Falstaff. There's no more faith in thee than in a stewed prune; nor no more truth in thee than in a drawn fox;
KING HENRY IV., Part I. Act iii. Scene 3.

Falstaff. To wake a wolf is as bad as to smell a fox.
KING HENRY IV., Part II. Act i. Scene 2.

Suffolk. The fox barks not when he would steal the lamb.

. . . .

. . . were't not madness then,
To make the fox surveyor of the fold?

. . . .

No; let him die, in that he is a fox,
KING HENRY VI., Part II. Act iii. Scene 1.

Gloster (*aside*). But when the fox hath once got in his nose,
He'll soon find means to make the body follow.
KING HENRY VI., Part III. Act iv. Scene 7.

Cressida. . . . when they have said,—as false
As air, as water, as wind, as sandy earth,

As fox to lamb, as wolf to heifer's calf,
Pard to the hind, or stepdame to her son ;—
Yea, let them say, to stick the heart of falsehood,
As false as Cressid.
TROILUS AND CRESSIDA, Act iii. Scene 2.

Marcius. . . . He that trusts to you,
Where he should find you lions finds you hares;
Where foxes, geese :
CORIOLANUS, Act i. Scene 1.

Timon. . . . If thou wert the lion, the fox would beguile thee; if thou wert the lamb, the fox would eat thee : if thou wert the fox, the lion would suspect thee, when, peradventure, thou wert accused by the ass :
TIMON OF ATHENS, Act iv. Scene 3.

Fool. A fox, when one has caught her,
    And such a daughter,
    Should sure to the slaughter,
If my cap would buy a halter;
So the fool follows after.
KING LEAR, Act i. Scene 4.

Or at the fox, which lives by subtlety,
VENUS AND ADONIS.

HARE.

Prince Henry. What say'st thou to a hare, . . . ?
KING HENRY IV., Part I. Act i. Scene 2.

Cressida. . . . They that have the voice of lions, and the act of hares, are they not monsters?
TROILUS AND CRESSIDA, Act iii. Scene 2.

Aufidius. If I fly, Marcius,
Halloo me like a hare.
Coriolanus, Act i. Scene 8.

Soldier. . . . the hare, the lion.
Macbeth, Act i. Scene 2.

But if thou needs will hunt, be ruled by me;
Uncouple at the timorous flying hare,

. . . .

" And when thou hast on foot the purblind hare,
　Mark the poor wretch, to overshoot his troubles,
　How he outruns the wind, and with what care
　He cranks and crosses, with a thousand doubles:
　　The many musits through the which he goes
　　Are like a labyrinth to amaze his foes.

" Sometime he runs among a flock of sheep,
　To make the cunning hounds mistake their smell,
　And sometime where earth-delving conies keep,
　To stop the loud pursuers in their yell;
　　And sometime sorteth with a herd of deer:
　　Danger deviseth shifts; wit waits on fear:

" For there his smell with others being mingled,
　The hot scent-snuffing hounds are driven to doubt,
　Ceasing their clamorous cry till they have singled
　With much ado the cold fault cleanly out;
　　Then do they spend their mouths: Echo replies,
　　As if another chase were in the skies.

" By this, poor Wat, far off upon a hill,
　Stands on his hinder legs with listening ear,

To hearken if his foes pursue him still;
Anon their loud alarms he doth hear;
   And now his grief may be comparèd well
   To one sore sick that hears the passing-bell.

"Then shalt thou see the dew-bedabbled wretch
   Turn, and return, indenting with the way;
   Each envious brier his weary legs doth scratch,
   Each shadow makes him stop, each murmur stay:
   For misery is trodden on by many,
   And being low never relieved by any.
VENUS AND ADONIS.

RABBIT.

Moth. . . . a rabbit on a spit;
LOVE'S LABOUR'S LOST, Act iii. Scene 1.

Biondello. . . . she went to the garden for parsley to stuff a rabbit;
TAMING OF THE SHREW, Act iv. Scene 4.

Falstaff. . . . hang me up by the heels for a rabbit sucker, or a poulter's hare.
KING HENRY IV., Part I. Act ii. Scene 4.

CONEY.

Orlando. Are you native of this place?
Rosalind. As the coney, that you see dwell where she is kindled.
AS YOU LIKE IT, Act iii. Scene 2.

Northumberland. So doth the coney struggle in the net.
>> King Henry VI., Part III. Act i. Scene 4.

3d Servant. . . . they will out of their burrows, like conies after rain,
>> Coriolanus, Act iv. Scene 5.

SQUIRREL.

Titania. I have a venturous fairy that shall seek
The squirrel's hoard, and fetch thee new nuts.
>> Midsummer Night's Dream, Act iv. Scene 1.

Mercutio. Her chariot is an empty hazel nut,
Made by the joiner squirrel,
>> Romeo and Juliet, Act i. Scene 4.

MOLE.

Caliban. Pray you, tread softly, that the blind mole may not
Hear a foot fall :
>> The Tempest, Act iv. Scene 1.

Autolycus. . . . I will bring these two moles, these blind ones, aboard him :
>> A Winter's Tale, Act iv. Scene 3.

Hamlet. Well said, old mole ! canst work i' the ground so fast ?
>> Hamlet, Act i. Scene 5.

Pericles.   The blind mole casts
Copp'd hills toward heaven, to tell, the earth is throng'd
By man's oppression ;
                    PERICLES, Act i. Scene 1.

WEASEL.

Jaques.   I can suck melancholy out of a song, as a weasel sucks eggs.
                    AS YOU LIKE IT, Act ii. Scene 5.

Lady. Out, you mad-headed ape !
A weasel hath not such a deal of spleen
As you are toss'd with.
                    KING HENRY IV., Part I. Act ii. Scene 3.

Westmoreland. For once the eagle England being in prey,
To her unguarded nest the weasel Scot
Comes sneaking, and so sucks her princely eggs ;
                    KING HENRY V., Act i. Scene 2.

Hamlet. Methinks, it is like a weasel.
Polonius. It is backed like a weasel.
                    HAMLET, Act iii. Scene 2.

Pisanio. As quarrellous as the weasel ;
                    CYMBELINE, Act iii. Scene 4.

FERRET.

Brutus. . . . such ferret and such fiery eyes,
                    JULIUS CÆSAR, Act i. Scene 2.

HEDGEHOG.

Caliban. . . . then like hedgehogs, which
Lie tumbling in my barefoot way, and mount
Their pricks at my footfall;
THE TEMPEST, Act ii. Scene 2.

1st Fairy. . . . . . .
Thorny hedgehogs, be not seen;
A MIDSUMMER NIGHT'S DREAM, Act ii. Scene 2.

Gloster. I grant ye.
Anne. Dost *grant* me, hedgehog?
KING RICHARD III., Act i. Scene 2.

2d Witch. Thrice; and once the hedge-pig whined.
MACBETH, Act iv. Scene 1.

CAT.

Antonio. They'll take suggestion as a cat laps milk;
THE TEMPEST, Act ii. Scene 1.

Claud. What! courage, man! What though care killed a cat, thou hast mettle enough in thee to kill care.
MUCH ADO ABOUT NOTHING, Act v. Scene 1.

Touchstone. If the cat will after kind,
So, be sure, will Rosalind.
AS YOU LIKE IT, Act iii. Scene 2.

Glendower. . . . at my birth,
The frame and huge foundation of the earth
Shaked like a coward.

Hotspur. Why, so it would have done at the same season, if your mother's cat had but kittened, though yourself had never been born.

Hotspur. . . . a ramping cat,

Falstaff. I am as vigilant as a cat to steal cream.
KING HENRY IV., Part I. Act iii. Scene 1;
Act iv. Scene 3.

Thersites. To be a dog, a mule, a cat, a fitchew, a toad, a lizard, an owl, a puttock, or a herring without a roe, I would not care: but to be Menelaus, I would conspire against destiny.
TROILUS AND CRESSIDA, Act v. Scene 1.

Iago. Come, be a man. Drown thyself? drown cats and blind puppies.
OTHELLO, Act i. Scene 3.

Marcius. The mouse ne'er shunn'd the cat as they did budge
From rascals worse than they.
CORIOLANUS, Act i. Scene 6.

Hamlet. Let Hercules himself do what he may,
The cat will mew, and dog will have his day.
HAMLET, Act v. Scene 1.

Benvolio. Why, what is Tybalt?
Mercutio. More than prince of cats, I can tell you.
ROMEO AND JULIET, Act ii. Scene 4.

Gower. The cat, with eyne of burning coal,
Now couches from the mouse's hole;
PERICLES, Act iii.

Edgar. Pur! the cat is gray.
KING LEAR, Act iii. Scene 6.

Lady M. Letting *I dare not* wait upon *I would,*
Like the poor cat i' the adage.

1st Witch. Thrice the brinded cat hath mew'd.
MACBETH, Act i. Scene 7; Act iv. Scene 1.

Shylock. Some, that are mad if they behold a cat;
MERCHANT OF VENICE, Act iv. Scene 1.

Hotspur. I had rather be a kitten, and cry *mew,*
KING HENRY IV., Part I. Act iii. Scene 1.

RAT.

Prospero. In few, they hurried us aboard a bark;
Bore us some leagues to sea; where they prepared
A rotten carcase of a boat, not rigg'd,
Nor tackle, sail, nor mast; the very rats
Instinctively have quit it;
TEMPEST, Act i. Scene 2.

Shallow. I have seen the time with my long sword I would have made you four tall fellows skip like rats.
MERRY WIVES OF WINDSOR, Act ii. Scene 1.

Claudio. Our natures do pursue

(Like rats that ravin down their proper bane)
A thirsty evil; and when we drink, we die.
MEASURE FOR MEASURE, Act i. Scene 3.

Rosalind. I was never so be-rhymed since Pythagoras' time, that I was an Irish rat, which I can hardly remember.
AS YOU LIKE IT, Act iii. Scene 2.

Shylock. What if my house be troubled with a rat, And I be pleased to give ten thousand ducats To have it baned?
MERCHANT OF VENICE, Act iv. Scene 1.

Menenius. Rome and her rats are at the point of battle,
The one side must have bale.—

Marcius. The Volsces have much corn; take these rats thither
To gnaw their garners.
CORIOLANUS, Act i. Scene 1.

Queen. Behind the arras hearing something stir, He whips his rapier out, and cries, *A rat! a rat!*
HAMLET, Act iv. Scene 1.

Mercutio. Tybalt, you rat-catcher, will you walk?
Tybalt. What wouldst thou have with me?
Mercutio. Good king of cats, nothing, but one of your nine lives;
ROMEO AND JULIET, Act iii. Scene 1.

Kent. Such smiling rogues as these,
Like rats, oft bite the holy cords a-twain
Which are too intrinse t' unloose:
KING LEAR, Act ii. Scene 2.

Cornelius. . . . *If Pisanio
Have,* said she, *given his mistress that confection
Which I gave him for cordial, she is served
As I would serve a rat.*
                    CYMBELINE, Act v. Scene 5.

1st *Witch.* Her husband's to Aleppo gone, master
    o' the Tiger :
But in a sieve I'll thither sail,
And, like a rat without a tail,
I'll do, I'll do, and I'll do.
                    MACBETH, Act i. Scene 3.

MOUSE.

Lucio. He—to give fear to use and liberty,
Which have, for long, run by the hideous law,
As mice by lions—
            MEASURE FOR MEASURE, Act i. Scene 5.

Lion. You, ladies, you, whose gentle hearts do fear
The smallest monstrous mouse that creeps on floor,

Puck. . . . not a mouse
Shall disturb this hallow'd house :
        MIDSUMMER NIGHT'S DREAM, Act v. Scenes 1 and 2.

Falstaff. Thou wilt be as valiant as the
wrathful dove, or most magnanimous mouse.—
            KING HENRY IV., Part II. Act iii. Scene 2.

Westmoreland. Playing the mouse, in absence of
    the cat,
To spoil and havoc more than she can eat.
                KING HENRY V., Act i. Scene 2.

Alençon. . . . piteous they will look, like drowned mice.
                KING HENRY VI., Part I. Act i. Scene 2.

Bernardo. Have you had quiet guard?
Francisco.   Not a mouse stirring.
                HAMLET, Act i. Scene 1.

Mercutio. What, a dog, a rat, a mouse, a cat, to scratch a man to death!

Romeo. . . . heaven is here,
Where Juliet lives; and every cat, and dog,
And little mouse, every unworthy thing,
Live here in heaven, and may look on her,
But Romeo may not.—
                ROMEO AND JULIET, Act iii. Scenes 1 and 3.

Edgar. But mice, and rats, and such small deer,
Have been Tom's food for seven long year.

Edgar. The fishermen, that walk upon the beach,
Appear like mice;
                KING LEAR, Act iii. Scene 4; Act iv. Scene 6.

ELEPHANT.

Decius. . . . he loves to hear
That unicorns may be betrayed with trees,
And bears with glasses, elephants with holes,
Lions with toils, and men with flatterers:
                JULIUS CÆSAR, Act ii. Scene 1.

Alexander. They say he is a very man *per se*,
And stands alone.
Cressida. So do all men; unless they are drunk, sick, or have no legs.
Alexander. This man, lady, hath robbed many beasts of their particular additions; he is as valiant as the lion, churlish as the bear, slow as the elephant:

Ulysses. The elephant hath joints, but none for courtesy:
TROILUS AND CRESSIDA, Act i. Scene 2;
Act ii. Scene 3.

CAMEL.

King Richard. It is as hard to come, as for a camel To thread the postern of a needle's eye.
KING RICHARD II., Act v. Scene 5.

Cressida. There is among the Greeks, Achilles; a better man than Troilus.
Pandarus. Achilles? a drayman, a porter, a very camel.
TROILUS AND CRESSIDA, Act i. Scene 2.

Brutus. . . . holding them,
In human action and capacity,
Of no more soul, nor fitness for the world,
Than camels in their war;
CORIOLANUS, Act ii. Scene 1.

Hamlet. Do you see that cloud, that's almost in shape like a camel?

Polonius. By the mass, and 'tis like a camel indeed.
HAMLET, Act iii. Scene 2.

LION.

Antonio. O, 'twas a din to fright a monster's ear;
To make an earthquake! sure, it was the roar
Of a whole herd of lions.
THE TEMPEST, Act ii. Scene 1.

Duke. Even like an o'ergrown lion in a cave,
That goes not out to prey.
MEASURE FOR MEASURE, Act i. Scene 4.

Quince. . . . Snug, the joiner, you, the lion's part:—and, I hope, here is a play fitted.
Snug. Have you the lion's part written? pray you, if it be, give it me, for I am slow of study.
Quince. You may do it extempore, for it is nothing but roaring.
MIDSUMMER NIGHT'S DREAM, Act i. Scene 2.

Holofernes. You have put me out of countenance.
Biron. False; we have given thee faces.
Holofernes. But you have out-faced them all.
Biron. An thou wert a lion, we would do so.
LOVE'S LABOUR'S LOST, Act v. Scene 2.

Prince Morocco. Yea, mock the lion when he roars for prey,

Jessica. In such a night,
Did Thisbe fearfully o'ertrip the dew;

And saw the lion's shadow ere himself,
And ran dismay'd away.
     MERCHANT OF VENICE, Act ii. Scene 1;
     Act v. Scene 1.

Petrucio. Have I not in my time heard lions roar?
     TAMING OF THE SHREW, Act i. Scene 2.

Bast. He that perforce robs lions of their hearts,
May easily win a woman's.

King Philip. Richard, that robb'd the lion of his heart,
And fought the holy wars in Palestine,

Bast. You are the hare of whom the proverb goes,
Whose valour plucks dead lions by the beard.

Bast. Talks as familiarly of roaring lions,
As maids of thirteen do of puppy-dogs!
     KING JOHN, Act i. Scene 1; Act ii.
     Scenes 1 and 2.

King Richard. . . . —lions make leopards tame.
     KING RICHARD II., Act i. Scene 1.

Falstaff. I shall think the better of myself, and thee, during my life; I for a valiant lion, and thou for a true prince.
     KING HENRY IV., Part I. Act ii. Scene 4.

King Henry. The man that once did sell the lion's skin
While the beast lived, was killed with hunting him.
     KING HENRY V., Act iv. Scene 3.

Talbot. Hark, countrymen! either renew the fight,
Or tear the lions out of England's coat;
>> KING HENRY VI., Part I. Act i. Scene 5.

Queen Margaret. Small curs are not regarded when
 they grin;
But great men tremble when the lion roars;
>> KING HENRY VI., Part II. Act iii. Scene 1.

Richard. Methought, he bore him in the thickest
 troop
As doth a lion in a herd of neat:
>> KING HENRY VI., Part III. Act ii. Scene 1.

Wolsey. He parted frowning from me, as if ruin
Leap'd from his eyes: so looks the chafèd lion
Upon the daring huntsman that has gall'd him;
>> KING HENRY VIII., Act iii. Scene 2.

Casca. Against the Capitol I met a lion,
Who glared upon me, and went surly by
Without annoying me:
>> JULIUS CÆSAR, Act i. Scene 3.

Enobarbus (aside). 'Tis better playing with a lion's
 whelp,
Than with an old one dying.
>> ANTONY AND CLEOPATRA, Act iii. Scene 11.

Patroclus. . . . . .
And, like a dew-drop from the lion's mane,
Be shook to airy air.
>> TROILUS AND CRESSIDA, Act iii. Scene 3.

Hamlet. My fate cries out,
And makes each petty artery in this body
As hardy as the Nemean lion's nerve.—
>> HAMLET, Act i. Scene 4.

Gent. This night, wherein the cub-drawn bear
  would couch,
The lion and the belly-pinched wolf
Keep their fur dry, unbonneted he runs,
And bids what will take all.
KING LEAR, Act iii. Scene 1.

Posthumus. . . . to grin like lions
Upon the pikes o' the hunters.
CYMBELINE, Act v. Scene 3.

Lavinia. Yet have I heard,—oh could I find it
  now!—
The lion, moved with pity, did endure
To have his princely paws pared all away.
TITUS ANDRONICUS, Act ii. Scene 3.

"To see his face the lion walk'd along
  Behind some hedge, because he would not fear him;
VENUS AND ADONIS.

Devouring Time, blunt thou the lion's paws,
SONNET XIX.

LIONESS.

Oliver.   .   .   .   .   .
A lioness, with udders all drawn dry,
Lay couching, head on ground, with cat-like watch,
AS YOU LIKE IT, Act iv. Scene 3.

Bast. . . . (*To Austria*) Sirrah, were I at
  home,
At your den, sirrah, with your lioness,

I'd set an ox-head to your lion's hide,
And make a monster of you.
KING JOHN, Act ii. Scene 1.

Calphurnia. A lioness hath whelped in the streets;
JULIUS CÆSAR, Act ii. Scene 2.

TIGER.

Proteus. For Orpheus' lute was strung with poets' sinews;
Whose golden touch could soften steel and stones,
Make tigers tame, and huge leviathans
Forsake unsounded deeps to dance on sands.
TWO GENTLEMEN OF VERONA, Act iii. Scene 2.

Helena. . . . the mild hind
Makes speed to catch the tiger: bootless speed!
When cowardice pursues, and valour flies.
MIDSUMMER NIGHT'S DREAM, Act ii. Scene 1.

York. O, tiger's heart, wrapp'd in a woman's hide!
. . . .
But you are more inhuman, more inexorable,
O, ten times more, than tigers of Hyrcania.
KING HENRY VI., Part III. Act i. Scene 4.

Nestor. The herd hath more annoyance by the brize
Than by the tiger;

Troilus. . . . when we vow to weep seas, live in fire, eat rocks, tame tigers;
TROILUS AND CRESSIDA, Act i. Scene 3;
Act iii. Scene 2.

Menenius. . . . there is no more mercy in him than there is milk in a male tiger;
CORIOLANUS, Act v. Scene 4.

Timon. . . . tigers, dragons, wolves, and bears;
TIMON OF ATHENS, Act iv. Scene 3.

Romeo. The time and my intents are savage-wild;
More fierce, and more inexorable far,
Than empty tigers, or the roaring sea.
ROMEO AND JULIET, Act v. Scene 3.

Albany. What have you done?
Tigers, not daughters, what have you perform'd?
KING LEAR, Act iv. Scene 2.

Lavinia. When did the tiger's young ones teach the dam?
TITUS ANDRONICUS, Act ii. Scene 3.

To recreate himself, when he hath sung,
The tiger would be tame, and gently hear him;
VENUS AND ADONIS.

Pluck the keen teeth from the fierce tiger's jaws.
SONNET XIX.

LEOPARD, LIBBARD, PARD.

Prospero. . . . and more pinch-spotted make them,
Than pard or cat o' mountain.
TEMPEST, Act iv. Scene 1.

Oberon. What thou seest, when thou dost wake,
Do it for thy true-love take;
Love and languish for his sake;
Be it ounce, or cat, or bear,
Pard, or boar with bristled hair,
   MIDSUMMER NIGHT'S DREAM, Act ii. Scene 2.

Costard. *I Pompey am,—*
Boyet.   With libbard's head on knee.
   LOVE'S LABOUR'S LOST, Act v. Scene 2.

Talbot. Sheep run not half so treacherous from the wolf,
Or horse, or oxen, from the leopard,
As you fly from your oft-subduèd slaves.
   KING HENRY VI., Part I. Act i. Scene 5.

Timon. . . . wert thou a horse, thou wouldst be seized by the leopard: wert thou a leopard, thou wert german to the lion, and the spots of thy kindred were jurors on thy life: all thy safety were remotion; and thy defence, absence.
   TIMON OF ATHENS, Act iv. Scene 3.

PANTHER.

Titus. To-morrow, an it please your majesty
To hunt the panther and the hart with me,
With horn and hound, we'll give your grace *bon-jour*.

Marcus.   I have dogs, my lord,
Will rouse the proudest panther in the chase,

Aaron. . . . . . .
Straight will I bring you to the loathsome pit,
Where I espied the panther fast asleep.
>> TITUS ANDRONICUS, Act i. Scene 2;
>> Act ii. Scenes 2 and 4.

HYENA.

Rosalind. . . . I will laugh like a hyen, and that when thou art inclined to sleep.
>> AS YOU LIKE IT, Act iv. Scene 1.

RHINOCEROS.

Macbeth. What man dare, I dare:
Approach thou like the rugged Russian bear,
The arm'd rhinoceros, or the Hyrcan tiger;
>> MACBETH, Act iii. Scene 4.

UNICORN.

Sebastian. Now I will believe
That there are unicorns;
>> THE TEMPEST, Act iii. Scene 3.

Decius. . . . he loves to hear
That unicorns may be betray'd with trees,
>> JULIUS CÆSAR, Act ii. Scene 1.

Timon. . . . wert thou the unicorn, pride and wrath would confound thee,
>> TIMON OF ATHENS, Act iv. Scene 3.

BEAR.

Dromio S. As from a bear a man would run for life,
So fly I from her that would be my wife.
     Comedy of Errors, Act iii. Scene 2.

Slender. . . . Why do your dogs bark so? be there bears i' the town?
Anne. I think there are, sir; I heard them talked of.
Slender. I love the sport well; but I shall as soon quarrel at it, as any man in England.—You are afraid, if you see the bear loose, are you not?
Anne. Ay, indeed, sir.
Slender. That's meat and drink to me, now: I have seen Sackerson loose twenty times; and have taken him by the chain: but, I warrant you, the women have so cried and shrieked at it, that it passed:—but women, indeed, cannot abide 'em; they are very ill-favoured rough things.
    Merry Wives of Windsor, Act i. Scene 1.

Oberon. The next thing then she waking looks upon,
Be it on lion, bear, or wolf, or bull,
 .   .   .   .
She shall pursue it with the soul of love.
    Midsummer Night's Dream, Act ii. Scene 1.

Claudio. . . . Hero and Margaret have by this played their parts with Beatrice; and then the two bears will not bite one another when they meet.
    Much Ado about Nothing, Act iii. Scene 2.

Prince of Morocco. Pluck the young sucking cubs from the she-bear,
>> MERCHANT OF VENICE, Act ii. Scene 1.

Clown. . . . And then for the land-service,—to see how the bear tore out his shoulder-bone; how he cried to me for help, and said his name was Antigonus, a nobleman.— . . . . how the poor gentleman roared, and the bear mocked him,
>> A WINTER'S TALE, Act iii. Scene 3.

Orleans. Foolish curs! that run winking into the mouth of a Russian bear,
>> KING HENRY V., Act iii. Scene 7.

York. Call hither to the stake my two brave bears,
That, with the very shaking of their chains,
They may astonish these fell lurking curs;
>> KING HENRY VI., Part II. Act v. Scene 1.

Antony. Sometime we see a cloud that's dragonish:
A vapour, sometime, like a bear, or lion,
>> ANTONY AND CLEOPATRA, Act iv. Scene 12.

Alexander. . . . churlish as the bear,
>> TROILUS AND CRESSIDA, Act i. Scene 2.

Othello. O, she will sing the savageness out of a bear!
>> OTHELLO, Act iv. Scene 1.

Brutus. He's a lamb, indeed, that baes like a bear.
Menenius. He's a bear, indeed, that lives like a lamb.
>> CORIOLANUS, Act ii. Scene 1.

Timon. . . . wert thou a bear, thou wouldst be killed by the horse;
>> TIMON OF ATHENS, Act iv. Scene 3.

Juliet. O, bid me leap, rather than marry Paris,
From off the battlements of yonder tower;
. . . . chain me with roaring bears;
>> ROMEO AND JULIET, Act iv. Scene 1.

Fool. . . . Horses are tied by the heads; dogs and bears by the neck;
>> KING LEAR, Act ii. Scene 4.

WOLF.

Don Pedro. Good morrow, masters; put your torches out:
The wolves have prey'd: and look, the gentle day,
Before the wheels of Phœbus, round about
Dapples the drowsy east with spots of gray;
>> MUCH ADO ABOUT NOTHING, Act v. Scene 3.

Gratiano. Thou almost mak'st me waver in my faith,
To hold opinion with Pythagoras,
That souls of animals infuse themselves
Into the trunks of men: thy currish spirit
Governed a wolf,
>> MERCHANT OF VENICE, Act iv. Scene 1.

Antigonus. —Come on, poor babe:
Some powerful spirit instruct the kites and ravens
To be thy nurses! Wolves and bears, they say,
Casting their savageness aside, have done
Like offices of pity.—
>> A WINTER'S TALE, Act ii. Scene 3.

Chief-Justice. . . . since all is well, keep it so: wake not a sleeping wolf.
    KING HENRY IV., Part II. Act i. Scene 2.

Constable. . . . they will eat like wolves,
    KING HENRY V., Act iii. Scene 7.

Shepherd. I wish some ravenous wolf had eaten thee!
    KING HENRY VI., Part I. Act v. Scene 4.

Gloster. Ah, thus king Henry throws away his crutch,
Before his legs be firm to bear his body:
Thus is the shepherd beaten from thy side,
And wolves are gnarling who shall gnaw thee first.
    KING HENRY VI., Part II. Act iii. Scene 1.

Richard. Nay, Warwick, single out some other chase;
For I myself will hunt this wolf to death.
    KING HENRY VI., Part III. Act ii. Scene 4.

Queen Elizabeth. Wilt thou, O God, fly from such gentle lambs,
And throw them in the entrails of the wolf?
    KING RICHARD III., Act iv. Scene 4.

Buckingham.   This holy fox,
Or wolf, or both (for he is equal ravenous
As he is subtle; and as prone to mischief,
As able to perform it: his mind and place
Infecting one another, yea, reciprocally,)
    KING HENRY VIII., Act i. Scene 1.

Cassius. And why should Cæsar be a tyrant, then?
Poor man! I know he would not be a wolf,
But that he sees the Romans are but sheep:
     JULIUS CÆSAR, Act i. Scene 3.

Ulysses. And appetite, an universal wolf,
So doubly seconded with will and power,
Must make, perforce, an universal prey,
And, last, eat up himself.
    TROILUS AND CRESSIDA, Act i. Scene 3.

Sicinius. Nature teaches beasts to know their friends.
Menenius. Pray you, who does the wolf love?
Sicinius. The lamb.
     CORIOLANUS, Act ii. Scene 1.

Lear. To be a comrade with the wolf and owl,—

Fool. He's mad that trusts in the tameness of a wolf,
   KING LEAR, Act ii. Scene 4; Act iii. Scene 6.

Arviragus. . . . subtle as the fox, for prey;
Like warlike as the wolf, for what we eat:
     CYMBELINE, Act iii. Scene 3.

Macbeth. . . . and wither'd murder,
Alarum'd by his sentinel, the wolf,
Whose howl's his watch,
    MACBETH, Act ii. Scene 1.

If he had spoke, the wolf would leave his prey,
And never fright the silly lamb that day.
     VENUS AND ADONIS.

PORCUPINE.

York. In Ireland have I seen this stubborn Cade
Oppose himself against a troop of kerns;
And fought so long, till that his thighs with darts
Were almost like a sharp-quill'd porcupine:
KING HENRY VI., Part II. Act iii. Scene I.

Ajax. Do not, porcupine, do not;
TROILUS AND CRESSIDA, Act ii. Scene I.

Ghost. . . . But that I am forbid
To tell the secrets of my prison-house,
I could a tale unfold, whose lightest word
Would harrow up thy soul; freeze thy young blood:
Make thy two eyes, like stars, start from their spheres;
Thy knotted and combinèd locks to part,
And each particular hair to stand an end,
Like quills upon the fretful porcupine;
HAMLET, Act i. Scene 5.

MONKEY.

Oberon. On meddling monkey, or on busy ape.
MIDSUMMER NIGHT'S DREAM, Act ii. Scene I.

Rosalind. . . . more giddy in my desires than a monkey:
AS YOU LIKE IT, Act iv. Scene I.

Tubal. One of them showed me a ring, that he had of your daughter for a monkey.

Shylock. Out upon her! Thou torturest me, Tubal: it was my turquoise: I had it of Leah, when I was a bachelor: I would not have given it for a wilderness of monkeys.
MERCHANT OF VENICE, Act iii. Scene 1.

Fool. . . . Horses are tied by the heads;
. . . . monkeys by the loins;
KING LEAR, Act ii. Scene 4.

Imogen. What makes your admiration?
Jachimo (aside). It cannot be i' the eye; for apes and monkeys,
'Twixt two such shes, would chatter this way and
Contemn with mows the other:
CYMBELINE, Act i. Scene 6.

APE.

Caliban. Sometimes like apes, that moe and chatter at me,
And after, bite me;
THE TEMPEST, Act ii. Scene 2.

Isabel. . . . but man, proud man!
Dress'd in a little brief authority,
Most ignorant of what he's most assured,
His glassy essence,—like an angry ape,
Plays such fantastic tricks before high heaven,
As make the angels weep:
MEASURE FOR MEASURE, Act ii. Scene 2.

Beatrice. . . . so deliver I up my apes, and away to saint Peter:

Don Pedro. What a pretty thing man is, when he goes in his doublet and hose, and leaves off his wit!

Claud. He is then a giant to an ape: but then is an ape a doctor to such a man.

MUCH ADO ABOUT NOTHING, Act ii. Scene 1;
Act v. Scene 1.

Jaques. Well then, if ever I thank any man I'll thank you: but that they call compliments is like the encounter of two dog-apes;

Rosalind. . . . more new-fangled than an ape;

AS YOU LIKE IT, Act ii. Scene 5;
Act iv. Scene 1.

Lady. What is it carries you away?
Hotspur. Why, my horse, my love, my horse.
Lady. Out, you mad-headed ape!

KING HENRY IV., Part I. Act ii. Scene 3.

York. Uncle, my brother mocks both you and me;
Because that I am little, like an ape,
He thinks that you should bear me on your shoulders.

KING RICHARD III., Act iii. Scene 1.

Antony. You show'd your teeth like apes,

JULIUS CÆSAR, Act v. Scene 1.

Marcius. You souls of geese
That bear the shapes of men, how have you run
From slaves that apes would beat!

CORIOLANUS, Act i. Scene 4.

Hamlet. . . . and, like the famous ape,
To try conclusions, in the basket creep,
And break your own neck down.

Hamlet. . . . he keeps them, like an ape doth nuts, in the corner of his jaw;
>> HAMLET, Act iii. Scene 4; Act iv. Scene 2.

BABOON.

Falstaff. . . . I have grated upon my good friends for three reprieves for you and your coach-fellow, Nym; or else you had looked through the grate, like a geminy of baboons.
>> THE MERRY WIVES OF WINDSOR, Act ii. Scene 2.

Doll. They say Poins hath a good wit.
Falstaff. He a good wit? hang him, baboon!
>> KING HENRY IV., Part II. Act ii. Scene 4.

Iago. Ere I would say I would drown myself for the love of a Guinea-hen, I would change my humanity with a baboon.
>> OTHELLO, Act i. Scene 3.

2d Witch. Cool it with a baboon's blood,
Then the charm is firm and good.
>> MACBETH, Act iv. Scene 1.

MARMOZET.

Caliban. . . and instruct thee how to snare the nimble marmozet;
>> THE TEMPEST, Act ii. Scene 2.

POLECAT, MUSKCAT, FITCHEW.

Evans. What is *fair*, William?
William. Pulcher.
Quickly. Polecats! there are fairer things than polecats, sure.
MERRY WIVES OF WINDSOR, Act iv. Scene 1.

Clown. Here is a pur of fortune's, sir, or of fortune's cat, (but not a musk-cat,) that has fallen into the unclean fish-pond of her displeasure, and as he says, is muddied withal:
ALL'S WELL THAT ENDS WELL, Act v. Scene 2.

Thersites. To be . . . a fitchew, . . . I would not care:
TROILUS AND CRESSIDA, Act v. Scene 1.

OTTER.

Hostess. Say, what beast, thou knave thou?
Falstaff. What beast? why, an otter!
Prince Henry. An otter, sir John! why an otter?
Falstaff. Why? she's neither fish nor flesh;
KING HENRY IV., Part I. Act iii. Scene 3.

FISH.

3d Fisher. I marvel how the fishes live in the sea.
1st Fisher. Why, as men do a-land; the great ones eat up the little ones.
PERICLES, Act ii. Scene 1.

SALMON.

Fluellen. There is a river in Macedon; and there is also moreover a river at Monmouth: it is called Wye, at Monmouth; but it is out of my prains what is the name of the other river; but 'tis all one, 'tis alike as my fingers is to my fingers, and there is salmons in both.
KING HENRY V., Act iv. Scene 7.

TROUT.

Maria. . . . here comes the trout that must be caught with tickling.
TWELFTH NIGHT, Act ii. Scene 5.

COD.

Iago. She that in wisdom never was so frail,
To change the cod's head for the salmon's tail;
OTHELLO, Act ii. Scene 1.

MACKEREL.

Falstaff. . . . you may buy land now as cheap as stinking mackerel.
KING HENRY IV., Part I. Act ii. Scene 4.

HERRING.

Caius. By gar, de herring is no dead so as I vill kill him.
MERRY WIVES OF WINDSOR, Act ii. Scene 3.

Sir Toby. . . . —a plague o' these pickle herrings!—
TWELFTH NIGHT, Act i. Scene 5.

Falstaff. Go thy ways, old Jack ; die when thou wilt, if manhood, good manhood, be not forgot upon the face of the earth, then am I a shotten herring.
KING HENRY IV., Part I. Act ii. Scene 4.

Cade. We John Cade, so termed of our supposed father,—
Dick (aside). Or rather, of stealing a cade of herrings.
KING HENRY VI., Part II. Act iv. Scene 2.

Thersites. To be . . . a herring without a roe, I would not care :
TROILUS AND CRESSIDA, Act v. Scene 1.

Benvolio. Here comes Romeo, here comes Romeo.
Mercutio. Without his roe, like a dried herring:—
O, flesh, flesh, how art thou fishified!
     ROMEO AND JULIET, Act ii. Scene 4.

PILCHARD.

Clown. . . . fools are as like husbands as pilchards are to herrings,
     TWELFTH NIGHT, Act iii. Scene 1.

GURNET.

Falstaff. If I be not ashamed of my soldiers I am a soused gurnet.
     KING HENRY IV., Part I. Act iv. Scene 2.

CONGER.

Doll. Hang yourself, you muddy conger, hang yourself!

Falstaff. . . . . . .
. . . . eats conger and fennel;
     KING HENRY IV., Part II. Act ii. Scene 4.

POOR JOHN.

Gregory. 'Tis well thou art not fish; if thou had'st, thou had'st been poor John.
     ROMEO AND JULIET, Act i. Scene 1.

Trinculo.   What have we here? A man or a fish? Dead or alive? A fish: he smells like a fish; a very ancient and fish-like smell; a kind of, not of the newest, Poor-John.

THE TEMPEST, Act ii. Scene 2.

EEL.

Moth. I will praise an eel with the same praise.
Armado. What? that an eel is ingenious?
Moth. That an eel is quick.

LOVE'S LABOUR'S LOST, Act i. Scene 2.

Petrucio. Or is the adder better than the eel, Because his painted skin contents the eye?

TAMING OF THE SHREW, Act iv. Scene 3.

Bast. My arms such eel-skins stuff'd,

KING JOHN, Act i. Scene 1.

Falstaff. . . . you might have trussed him, and all his apparel, into an eel-skin;

KING HENRY IV., Part II. Act iii. Scene 2.

Fool. Cry to it, nuncle, as the cockney did to the eels, when she put 'em i' the paste alive; she knapp'd 'em o' the coxcombs with a stick, and cried, *Down, wantons, down:* 'twas her brother that, in pure kindness to his horse, butter'd his hay.

KING LEAR, Act ii. Scene 4.

PIKE. DACE.

Falstaff. I the young dace be a bait for the old

pike, I see no reason, in the law of nature, but I may snap at him.
>> KING HENRY IV., Part II. Act iii. Scene 2.

GUDGEON.

Gratiano. . . . fish not with this melancholy bait,
For this fool-gudgeon, this opinion.
>> MERCHANT OF VENICE, Act i. Scene 1.

TENCH.

2d Carrier. I think, this is the most villanous house in all London road for fleas: I am stung like a tench.
1st Carrier. Like a tench? by the mass, there is ne'er a king in Christendom could be better bit than I have been since the first cock.
>> KING HENRY IV., Part I. Act ii. Scene 1.

CARP.

Clown. . . . pray you, sir, use the carp as you may;
>> ALL'S WELL THAT ENDS WELL, Act v. Scene 2.

ANCHOVY.

Poins. Item, Anchovies and sack after supper, 2s. 6d.
>> KING HENRY IV., Part I. Act ii. Scene 4.

OYSTER.

Falstaff. I will not lend thee a penny.
Pistol. Why, then the world's mine oyster,
Which I with sword will open.
>> MERRY WIVES OF WINDSOR, Act ii. Scene 2.

Benedick. . . . I will not be sworn but love may transform me to an oyster; but I'll take my oath on it, till he have made an oyster of me, he shall never make me such a fool.
>> MUCH ADO ABOUT NOTHING, Act ii. Scene 3.

Touchstone. . . . Rich honesty dwells like a miser, sir, in a poor house; as your pearl in your foul oyster.
>> AS YOU LIKE IT, Act v. Scene 4.

King Richard. Off goes his bonnet to an oyster-wench;
>> KING RICHARD II., Act. i. Scene 4.

Alexas. *Good friend,* quoth he,
Say,—*The firm Roman to great Egypt sends
This treasure of an oyster;*
>> ANTONY AND CLEOPATRA, Act i. Scene 5.

Fool. Canst tell how an oyster makes his shell?
Lear. No.
Fool. Nor I neither;
>> KING LEAR, Act i. Scene 5.

CRAB.

Hamlet. . . . for you yourself, sir, should be old as I am, if, like a crab, you could go backward.
HAMLET, Act ii. Scene 2.

PRAWNS.

Hostess. Did not goodwife Keech, the butcher's wife, come in then, and call me gossip Quickly? coming in to borrow a mess of vinegar; telling us she had a good dish of prawns; whereby thou didst desire to eat some;
KING HENRY IV., Part II. Act ii. Scene 1.

SHRIMP.

Holofernes. Great Hercules is presented by this imp,
Whose club kill'd Cerberus, that three-headed canus;
And, when he was a babe, a child, a shrimp,
Thus did he strangle serpents in his manus :
LOVE'S LABOUR'S LOST, Act v. Scene 2.

Countess. Is this the scourge of France?
Is this the Talbot, so much fear'd abroad,
. . . .
Alas! this is a child, a silly dwarf:
It cannot be this weak and writhled shrimp
Should strike such terror to his enemies.
KING HENRY VI., Part I. Act ii. Scene 3.

MUSSEL.

Prospero. Sea-water shalt thou drink, thy food shall be
The fresh-brook mussels,
     THE TEMPEST, Act i. Scene 2.

Simple. Pray you, sir, was't not the wise woman of Brentford?
Falstaff. Ay, marry, was it, mussel-shell:
    MERRY WIVES OF WINDSOR, Act iv. Scene 5.

COCKLE.

Petrucio. Why, 'tis a cockle, or a walnut-shell,
    TAMING OF THE SHREW, Act iv. Scene 3.

Gower. Thus time we waste, and longest leagues make short,
Sail seas in cockles, have, an wish but for't;
    PERICLES, Act iv. GOWER.

BARNACLE.

Caliban. . . . we shall lose our time,
And all be turn'd to barnacles, or to apes
With foreheads villanous low.
    THE TEMPEST, Act iv. Scene 1.

WHALE.

Biron. To show his teeth as white as whale's bone:
    LOVE'S LABOUR'S LOST, Act v. Scene 2.

Nestor. And there they fly, or die, like scaled sculls
Before the belching whale;
    TROILUS AND CRESSIDA, Act v. Scene 5.

Hamlet. Or, like a whale?
Polonius. Very like a whale.
    HAMLET, Act iii. Scene 2.

1st Fisher.   I can compare our rich misers to nothing so fitly as to a whale; 'a plays and tumbles, driving the poor fry before him, and at last devours them all at a mouthful. Such whales have I heard on o' the land, who never leave gaping, till they've swallowed the whole parish, church, steeple, bells, and all.

Pericles. . . . nor have I time
To give thee hallow'd to thy grave, but straight
Must cast thee, scarcely coffin'd, in the ooze;
Where, for a monument upon thy bones,
And aye-remaining lamps, the belching whale
And humming water must o'erwhelm thy corpse,
Lying with simple shells.
    PERICLES, Act ii. Scene 1; Act iii. Scene 1.

SHARK.

3d Witch. . . . maw, and gulf,
Of the ravin'd salt-sea shark;
    MACBETH, Act iv. Scene 1.

PORPUS.

3d Fisher. Nay, master, said not I as much, when

I saw the porpus how he bounced and tumbled?
they say, they are half fish, half flesh; a plague on
them! they ne'er come but I look to be washed.
PERICLES, Act ii. Scene 1.

DOLPHIN. DOGFISH.

Talbot. Pucelle or puzzel, dolphin or dogfish,
Your hearts I'll stamp out with my horse's heels,
KING HENRY VI., Part I. Act i. Scene 4.

MINNOW.

Coriolanus. . . . . .
Hear you this Triton of the minnows?
CORIOLANUS, Act iii. Scene 1.

REPTILES.

Tamora. The snake lies rolled in the cheerful sun;
                            TITUS ANDRONICUS, Act ii. Scene 3.

CROCODILE.

Queen Margaret. Free lords, cold snow melts with
    the sun's hot beams.
Henry my lord is cold in great affairs,
Too full of foolish pity : and Gloster's show
Beguiles him, as the mournful crocodile
With sorrow snares relenting passengers;
           KING HENRY VI., Part II. Act iii. Scene 1.

Lepidus. Your serpent of Egypt is bred now of your mud by the operation of your sun : so is your crocodile.
  Antony. They are so.

    .   .   .   .

  Lepidus. What manner o' thing is your crocodile?
  Antony. It is shaped, sir, like itself; and it is as broad as it hath breadth : it is just so high as it is, and moves with its own organs : it lives by that which

nourisheth it: and the elements once out of it, it transmigrates.

Lepidus. What colour is it of?
Antony. Of its own colour too.
Lepidus. "Tis a strange serpent.
Antony. "Tis so. And the tears of it are wet.
Antony and Cleopatra, Act ii. Scene 7.

Othello. If that the earth could teem with woman's tears,
Each drop she falls would prove a crocodile:—
Othello, Act iv. Scene 1.

Hamlet. Woul't drink up Esil? eat a crocodile?
Hamlet, Act v. Scene 1.

TORTOISE.

Prospero. Come forth, I say; there's other business for thee:
Come, thou tortoise!
The Tempest, Act i. Scene 2.

Romeo. . . . . . .
And in his needy shop a tortoise hung,
Romeo and Juliet, Act v. Scene 1.

ALLIGATOR.

Romeo. An alligator stuff'd and other skins
Of ill-shaped fishes;
Romeo and Juliet, Act v. Scene 1.

CAMELEON.

Speed. Why muse you, sir? 'tis dinner-time.
Valentine. I have dined.
Speed. Ay, but hearken, sir; though the cameleon Love can feed on the air, I am one that am nourished by my victuals, and would fain have meat.
Two Gentlemen of Verona, Act ii. Scene 1.

Gloster. I can add colours to the cameleon;
King Henry VI., Part III. Act iii. Scene 2.

King. How fares our cousin Hamlet?
Hamlet. Excellent, i' faith; of the cameleon's dish: I eat the air, promise-crammed:
Hamlet, Act iii. Scene 2.

LIZARD.

Suffolk. Their softest touch as smart as lizards' stings!
King Henry VI., Part II. Act iii. Scene 2.

Queen Margaret. Mark'd by the destinies to be avoided,
As venom toads, or lizards' dreadful stings.
King Henry VI., Part III. Act ii. Scene 2.

2d Witch. Lizard's leg, and owlet's wing,
Macbeth, Act iv. Scene 1.

BAT. REREMOUSE.

Ariel. On the bat's back I do fly
After summer merrily.
THE TEMPEST, Act v. Scene 1.

Titania. Some, war with rear-mice for their leathern
wings,
To make my small elves coats;
MIDSUMMER NIGHT'S DREAM, Act ii. Scene 2.

Macbeth. . . . ere the bat hath flown
His cloister'd flight;

2d Witch. Wool of bat, and tongue of dog,
MACBETH, Act iii. Scene 2; Act iv. Scene 1.

TOAD.

Caliban. All the charms
Of Sycorax, toads, beetles, bats, light on you!
THE TEMPEST, Act i. Scene 2.

Duke S. Sweet are the uses of adversity;
Which, like the toad, ugly and venomous,
Wears yet a precious jewel in his head;
AS YOU LIKE IT, Act ii. Scene 1.

King Richard. . . . . .
And heavy-gaited toads, lie in their way;
KING RICHARD II., Act iii. Scene 2.

Anne. Never hung poison on a fouler toad.

Queen Elizabeth. O, thou didst prophesy the time would come
That I should wish for thee to help me curse
That bottled spider, that foul bunch-back'd toad.

Duchess. Thou toad, thou toad, where is thy brother Clarence?
And little Ned Plantagenet, his son?
KING RICHARD III., Act i. Scene 2;
Act iv. Scene 4.

Othello. . . . . .
Or keep it as a cistern, for foul toads
OTHELLO, Act iv. Scene 2.

Nurse. . . . but she, good soul, had as lieve see a toad, a very toad, as see him.

Juliet. Some say, the lark and loathèd toad change eyes;
O, now I would they had changed voices too!
ROMEO AND JULIET, Act ii. Scene 4;
Act iii. Scene 5.

Guiderius. Cloten, thou double villain, be thy name,
I cannot tremble at it; were't toad, or adder, spider,
'Twould move me sooner.
CYMBELINE, Act iv. Scene 2.

1st Witch. Toad, that under cold stone,
Days and nights hast thirty-one.
MACBETH, Act iv. Scene 1.

PADDOCK.

Hamlet. 'Twere good you let him know:
For who, that's but a queen, fair, sober, wise,
Would from a paddock, from a bat, a gib,
Such dear concernings hide?
                HAMLET, Act iii. Scene 4.

FROG.

Edgar. Poor Tom; that eats the swimming frog,
the toad, the tadpole, the wall-newt, and the water;
                KING LEAR, Act iii. Scene 4.

2d Witch. Eye of newt, and toe of frog,
                MACBETH, Act iv. Scene 1.

SERPENT.

Lysander. . . . vile thing, let loose;
Or I will shake thee from me, like a serpent.
                MIDSUMMER NIGHT'S DREAM, Act iii. Scene 2.

Antonio. . . . I loved my niece;
And she is dead, slander'd to death by villains;
That dare as well answer a man, indeed,
As I dare take a serpent by the tongue:
                MUCH ADO ABOUT NOTHING, Act v. Scene 1.

Countess. When I said, a mother,
Methought you saw a serpent:
                ALL'S WELL THAT ENDS WELL, Act i. Scene 3.

King John. He is a very serpent in my way;
KING JOHN, Act iii. Scene 3.

King Henry. Their touch affrights me as a serpent's sting.   .   .   .   .
  Salisbury. Were there a serpent seen, with forked tongue,
That slily glided towards your majesty,
It were but necessary you were waked;
Lest, being suffer'd in that harmful slumber,
The mortal worm might make the sleep eternal:
And therefore do they cry, though you forbid,
That they will guard you, whether you will or no,
From such fell serpents as false Suffolk is;
With whose envenomed and fatal sting,
Your loving uncle, twenty times his worth,
They say, is shamefully bereft of life.
  .   .   .   .
  Suffolk. Their music frightful as the serpent's hiss;
KING HENRY VI., Part II. Act iii. Scene 2.

Clifford. Who 'scapes the lurking serpent's mortal sting?
Not he that sets his foot upon her back.
KING HENRY VI., Part III. Act ii. Scene 2.

Brutus. And therefore think him as a serpent's egg,
Which, hatch'd, would as his kind grow mischievous;
And kill him in the shell.
JULIUS CÆSAR, Act ii. Scene 1.

Cleopatra.   He's speaking now,
Or murmuring, *Where's my serpent of old Nile?*
For so he calls me:

Cleopatra. . . . if knife, drugs, serpents, have edge, sting, or operation, I am safe:
Antony and Cleopatra, Act i. Scene 5;
Act iv. Scene 13.

Thersites. I will no more trust him when he leers, than I will a serpent when he hisses:
Troilus and Cressida, Act v. Scene 1.

Emilia. If any wretch have put this in your head, Let heaven requite it with the serpent's curse!
Othello, Act iv. Scene 2.

Aufidius. Not Afric owns a serpent I abhor More than thy fame, and envy:
Coriolanus, Act i. Scene 8.

Ghost. 'Tis given out, that, sleeping in mine orchard,
A serpent stung me;

. . . .

. . . . but know, thou noble youth,
The serpent that did sting thy father's life,
Now wears his crown.
Hamlet, Act i. Scene 5.

Lear. . . . sharper than a serpent's tooth it is
To have a thankless child.
King Lear, Act i. Scene 4.

Lady Macbeth. . . . bear welcome in your eye, Your hand, your tongue: look like the innocent flower,
But be the serpent under it.
Macbeth, Act i. Scene 5.

ADDER.

Caliban. . . . sometime am I
All wound with adders, who, with cloven tongues,
Do hiss me into madness :—
     THE TEMPEST, Act ii. Scene 2.

Hermia. . . . O, brave touch !
Could not a worm, an adder, do so much?
An adder did it : for with doubler tongue
Than thine, thou serpent, never adder stung.
     MIDSUMMER NIGHT'S DREAM, Act iii. Scene 2.

King Richard. And when they from thy bosom
  pluck a flower,
Guard it, I pray thee, with a lurking adder,
Whose double tongue may with a mortal touch
Throw death upon thy sovereign's enemies.—
     KING RICHARD II., Act iii. Scene 2.

Queen Margaret. What! art thou, like the adder,
waxen deaf?
     KING HENRY VI., Part II. Act iii. Scene 2.

York. She wolf of France, but worse than wolves of
  France,
Whose tongue more poisons than the adder's tooth !
     KING HENRY VI., Part III. Act i. Scene 4.

Anne. More direful hap betide that hated wretch,
That makes us wretched by the death of thee,
Than I can wish to adders, spiders, toads,
Or any creeping venom'd thing that lives !
     KING RICHARD III., Act i. Scene 2.

Brutus. It is the bright day that brings forth the adder ;
JULIUS CÆSAR, Act ii. Scene 1.

Hector. . . . for pleasure, and revenge,
Have ears more deaf than adders to the voice
Of any true decision.
TROILUS AND CRESSIDA, Act ii. Scene 2.

Hamlet. There's letters seal'd : and my two school-fellows,—
Whom I will trust, as I will adders fang'd,—
They bear the mandate ;
HAMLET, Act iii. Scene 4.

Edmund. To both these sisters have I sworn my love ;
Each jealous of the other, as the stung
Are of the adder.
KING LEAR, Act v. Scene 1.

2d Witch. Adder's fork, and blind-worm's sting,
MACBETH, Act iv. Scene 1.

SNAKE.

Oberon. And there the snake throws her enamell'd skin,

1st Fairy. You spotted snakes, with double tongue,
MIDSUMMER NIGHT'S DREAM, Act ii.
Scenes 1 and 2.

Moth. . . . if any of the audience hiss, you

may cry, *Well done, Hercules! now thou crushest the snake!* that is the way to make an offence gracious; though few have the grace to do it.
                      LOVE'S LABOUR'S LOST, Act v. Scene 1.

Rosalind.   I see, love hath made thee a tame snake, . . .

. . . .

Oliver. A wretched ragged man, o'ergrown with hair,
Lay sleeping on his back : about his neck
A green and gilded snake had wreathed itself,
Who with her head, nimble in threats, approach'd
The opening of his mouth; but suddenly
Seeing Orlando, it unlink'd itself,
And with indented glides did slip away
Into a bush:
                      AS YOU LIKE IT, Act iv. Scene 3.

King Richard. Snakes, in my heart-blood warm'd, that sting my heart!
                      KING RICHARD II., Act iii. Scene 2.

Queen Margaret. Or as the snake, roll'd in a flowering bank,
With shining checker'd slough,

. . . .

York. I fear me you but warm the starvèd snake,
Who, cherish'd in your breasts, will sting your hearts.
                  KING HENRY VI., Part II. Act iii. Scene 1.

Cleopatra. Thou shouldst come like a Fury crown'd with snakes,
                ANTONY AND CLEOPATRA, Act ii. Scene 5.

Macbeth. We have scotch'd the snake, not kill'd it;

2d Witch. Fillet of a fenny snake,
        MACBETH, Act iii. Scene 2; Act iv. Scene 1.

SCORPION.

Queen Margaret.   Seek not a scorpion's nest,
        KING HENRY VI., Part II. Act iii. Scene 2.

Cornelius. Your daughter, whom she bore in hand to love
With such integrity, she did confess
Was as a scorpion to her sight;
        CYMBELINE, Act v. Scene 5.

Macbeth. O, full of scorpions is my mind, dear wife!
        MACBETH, Act iii. Scene 2.

VIPER.

Pistol.   O viper vile!
        KING HENRY V., Act ii. Scene 1.

Pandarus.   Why, they are vipers:
        TROILUS AND CRESSIDA, Act iii. Scene 1.

Lodovico. Where is that viper?
        OTHELLO, Act v. Scene 2.

NEWT.

Timon. The gilded newt, and eyeless venom'd worm,
        TIMON OF ATHENS, Act iv. Scene 3.

BLIND-WORM.

1st Fairy. Newts, and blind-worms, do no wrong:
        MIDSUMMER NIGHT'S DREAM, Act ii. Scene 2.

WORM.

Rosalind. . . . men have died from time to time, and worms have eaten them, but not for love.
        AS YOU LIKE IT, Act iv. Scene 1.

Katharine. Come, come, you froward and unable worms!
        TAMING OF THE SHREW, Act v. Scene 2.

Constance. And ring these fingers with thy household worms;
        KING JOHN, Act iii. Scene 4.

King Richard. Let's talk of graves, of worms, and epitaphs;
        KING RICHARD II., Act iii. Scene 2.

Exeter. 'Tis no sinister nor no awkward claim, Pick'd from the worm-holes of long-vanish'd days,
        KING HENRY V., Act ii. Scene 4.

King Henry. Civil dissension is a viperous worm
KING HENRY VI., Part I. Act iii. Scene 1.

Clifford. The smallest worm will turn being trodden on ;
And doves will peck in safeguard of their brood.
KING HENRY VI., Part III. Act ii. Scene 2.

Katharine. When I shall dwell with worms, and my poor name
Banish'd the kingdom !
KING HENRY VIII., Act iv. Scene 2.

Othello. The worms were hallow'd that did breed the silk ;
OTHELLO, Act iii. Scene 4.

King. Now, Hamlet, where's Polonius?
Hamlet. At supper.
King. At supper? Where?
Hamlet. Not where he eats, but where he is eaten : a certain convocation of politic worms are e'en at him. Your worm is your only emperor for diet : we fat all creatures else, to fat us ; and we fat ourselves for maggots : your fat king, and your lean beggar, is but variable service ; two dishes but to one table ; that's the end.
King. Alas, alas !
Hamlet. A man may fish with the worm that hath eat of a king ; and eat of the fish that hath fed of that worm.
HAMLET, Act iv. Scene 3.

Romeo. . . here, here will I remain
With worms
ROMEO AND JULIET, Act v. Scene 3.

Lear. Thou owest the worm no silk,
> KING LEAR, Act iii. Scene 4.

Pisanio. No, 'tis slander,—
Whose edge is sharper than the sword ; whose tongue
Outvenoms all the worms of Nile ;
> CYMBELINE, Act iii. Scene 4.

Macbeth. There the grown serpent lies ; the worm, that's fled,
Hath nature that in time will venom breed ;
> MACBETH, Act iii. Scene 4.

ASP.

Cleopatra. Hast thou the pretty worm of Nilus there,
That kills and pains not ?
Clown. Truly, I have him :

. . . .

Cleopatra. Have I the aspic in my lips ?

. . . .

1st Guard. This is an aspic's trail : and these fig-leaves
Have slime upon them, such as the aspic leaves
Upon the caves of Nile.
> ANTONY AND CLEOPATRA, Act v. Scene 2.

Othello. Yield up, O love, thy crown, and hearted throne,
To tyrannous hate ! swell, bosom, with thy fraught,
For 'tis of aspics' tongues !
> OTHELLO, Act iii. Scene 3.

SNAIL.

Luciana. . . . thou snail,
    COMEDY OF ERRORS, Act ii. Scene 2.

1st Fairy. Worm, nor snail, do no offence.
    MIDSUMMER NIGHT'S DREAM, Act ii. Scene 2.

Biron. Love's feeling is more soft, and sensible,
Than are the tender horns of cockled snails,
    LOVE'S LABOUR'S LOST, Act iv. Scene 3.

Jaques. . . . . . .
Then the whining schoolboy, with his satchel,
And shining morning face, creeping like snail
Unwillingly to school :

Rosalind. Nay, an you be so tardy, come no more in my sight ; I had as lief be wooed of a snail.
Orlando. Of a snail?
Rosalind. Ay, of a snail ; for though he comes slowly, he carries his house on his head ; a better jointure, I think, than you make a woman : besides, he brings his destiny with him.
Orlando. What's that?
Rosalind. Why, horns ;
    AS YOU LIKE IT, Act ii. Scene 7 ;
    Act iv. Scene 1.

Nestor. And bid the snail-paced Ajax arm for shame.—
    TROILUS AND CRESSIDA, Act v. Scene 5.

Fool. . . . I can tell why a snail has a house.
Lear. Why?
Fool. Why, to put his head in; not to give it away to his daughters, and leave his horns without a case.

KING LEAR, Act i. Scene 5.

SLUG.

Luciana. . . . thou slug,

COMEDY OF ERRORS, Act ii. Scene 2.

Prince. . . . what a slug is Hastings!

KING RICHARD III., Act iii. Scene 1.

INSECTS.

Mercutio. O, then, I see, Queen Mab hath been with you.
. . . .
Drawn with a team of little atomies
ROMEO AND JULIET, Act i. Scene 4.

BEE.

Prospero. . . . thou shalt be pinch'd
As thick as honeycomb, each pinch more stinging
Than bees that made them.

Ariel. Where the bee sucks, there suck I ;
THE TEMPEST, Act i. Scene 2 ; Act v. Scene 1.

King Henry. When, like the bee, culling from every flower
The virtuous sweets ;
Our thighs pack'd with wax, our mouths with honey,
We bring it to the hive ; and, like the bees,
Are murder'd for our pains.
KING HENRY IV., Part II. Act iv. Scene 4.

Canterbury. . . . for so work the honey-bees ;
Creatures, that, by a rule in nature, teach
The act of order to a peopled kingdom.
KING HENRY V., Act i. Scene 2.

Talbot. So bees with smoke, and doves with noisome stench,
Are from their hives and houses driven away.
    KING HENRY VI., Part I. Act i. Scene 5.

Warwick. The commons, like an angry hive of bees,
That want their leader, scatter up and down,
And care not who they sting in his revenge.

Cade. Some say the bee stings; but I say 't is the bee's wax, for I did but seal once to a thing, and I was never mine own man since.
    KING HENRY VI., Part II. Act iii. Scene 2;
    Act iv. Scene 2.

Cassius. . . . they rob the Hybla bees,
And leave them honeyless.
    JULIUS CÆSAR, Act v. Scene 1.

3d Fisher. We would purge the land of these drones, that rob the bee of her honey.
    PERICLES, Act ii. Scene 1.

Imogen. . . Good wax, thy leave:—bless'd be
You bees that make these locks of counsel!
    CYMBELINE, Act iii. Scene 2.

HUMBLE BEE.

Titania. The honey-bags steal from the humble-bees,
And, for night-tapers, crop their waxen thighs,

Bottom. Monsieur Cobweb; good monsieur, get your weapons in your hand, and kill me a red-hipped humble-bee on the top of a thistle; and, good monsieur, bring me the honey-bag. Do not fret yourself too much in the action, monsieur; and, good monsieur, have a care the honey-bag break not; I would be loth to have you overflown with a honey-bag, signior.
MIDSUMMER NIGHT'S DREAM, Act iii. Scene 1;
Act iv. Scene 1.

Armado. The fox, the ape, and the humble-bee,
Were still at odds, being but three.
LOVE'S LABOUR'S LOST, Act iii. Scene 1.

Lafeu. . . . red-tailed humble-bee
ALL'S WELL THAT ENDS WELL, Act iv. Scene 5.

Pandarus. Full merrily the humble-bee doth sing,
Till he hath lost his honey and his sting:
And being once subdued in armed tail,
Sweet honey and sweet notes together fail.—
TROILUS AND CRESSIDA, Act v. Scene 11.

DRONE.

Luciana. Dromio, thou drone,
COMEDY OF ERRORS, Act ii. Scene 2.

Shylock. . . . drones hive not with me,
MERCHANT OF VENICE, Act ii. Scene 5.

Canterbury. The lazy yawning drone.
KING HENRY V., Act i. Scene 2.

Suffolk. Drones suck not eagles' blood, but rob bee-hives :
>KING HENRY VI., Part II. Act iv. Scene 1.

Gower. Good Helicane hath stay'd at home,
Not to eat honey, like a drone,
From other's labours ;
>PERICLES, Act ii.   GOWER.

WASP.

Autolycus. He has a son, who shall be flayed alive ; then, 'nointed over with honey, set on the head of a wasps' nest ;
>A WINTER'S TALE, Act iv. Scene 3.

Petrucio. Come, come, you wasp ; i' faith, you are too angry.
Katharine. If I be waspish, best beware my sting.
Petrucio. My remedy is then, to pluck it out.
Katharine. Ay, if the fool could find out where it lies.
Petrucio. Who knows not where a wasp does wear his sting?
In his tail.
Katharine.   In his tongue.
>TAMING OF THE SHREW, Act ii. Scene 1.

Suffolk. There be more wasps that buzz about his nose,
Will make this sting the sooner.
>KING HENRY VIII., Act iii. Scene 2.

ANT.

Hotspur. . . . sometime he angers me,
With telling me of the moldwarp and the ant,
>> KING HENRY IV., Part I. Act iii. Scene 1.

Fool. We'll set thee to school to an ant, to teach thee there's no labouring i' the winter.
>> KING LEAR, Act ii. Scene 4.

BUTTERFLY.

Titania. And pluck the wings from painted butterflies,
To fan the moonbeams from his sleeping eyes:
>> MIDSUMMER NIGHT'S DREAM, Act iii. Scene 1.

Achilles. . . . for men, like butterflies,
Show not their mealy wings but to the summer;
>> TROILUS AND CRESSIDA, Act iii. Scene 3.

Valeria. I saw him run after a gilded butterfly; and when he caught it, he let it go again; and after it again; and over and over he comes, and up again; catched it again: or whether his fall enraged him, or how 't was, he did so set his teeth, and tear it; O, I warrant, how he mammocked it!

Cominius. . . . with no less confidence
Than boys pursuing summer butterflies,

Menenius. There is difference between a grub and a butterfly; yet your butterfly was a grub.
>> CORIOLANUS, Act i. Scene 3; Act iv. Scene 6;
>> Act v. Scene 4.

Lear. So we'll live,
And pray, and sing, and tell old tales, and laugh
At gilded butterflies,
     KING LEAR, Act v. Scene 3.

MOTH.

Portia. Thus hath the candle singed the moth.
     MERCHANT OF VENICE, Act ii. Scene 9.

Desdemona.   A moth of peace,
     OTHELLO, Act i. Scene 3.

Valeria. You would be another Penelope : yet, they say, all the yarn she spun in Ulysses' absence did but fill Ithaca full of moths.
     CORIOLANUS, Act i. Scene 3.

GRASSHOPPER.

Mercutio.
Her waggon-spokes made of long spinners' legs,
The cover of the wings of grasshoppers ;
     ROMEO AND JULIET, Act i. Scene 4.

GLOW-WORM.

Evans. And twenty glow-worms shall our lanterns be,
To guide our measure round about the tree.
     MERRY WIVES OF WINDSOR, Act v. Scene 5.

Titania. . . . . . .
And light them at the fiery glow-worm's eyes,
MIDSUMMER NIGHT'S DREAM, Act iii. Scene 1.

Ghost. The glow-worm shows the matin to be near,
And 'gins to pale his uneffectual fire :
HAMLET, Act i. Scene 5.

Pericles. . . . like a glow-worm in the night,
The which hath fire in darkness, none in light ;
PERICLES, Act ii. Scene 3.

His eyes like glow-worms shine when he doth fret :
VENUS AND ADONIS.

SPIDER.

Duke. To draw with idle spiders' strings
Most ponderous and substantial things :
MEASURE FOR MEASURE, Act iii. Scene 2.

1st Fairy. Weaving spiders, come not here :
Hence, you long-legg'd spinners, hence :
MIDSUMMER NIGHT'S DREAM, Act ii. Scene 2.

Bassanio. Here in her hairs
The painter plays the spider ; and hath woven
A golden mesh to entrap the hearts of men,
MERCHANT OF VENICE, Act iii. Scene 2.

Leontes. . . . There may be in the cup
A spider steep'd, and one may drink, depart,
And yet partake no venom ;
A WINTER'S TALE, Act ii. Scene 1.

Bast. . . . the smallest thread
That ever spider twisted
KING JOHN, Act iv. Scene 3.

King Richard. But let thy spiders, that suck up thy venom,
And heavy-gaited toads, lie in their way;
KING RICHARD II., Act iii. Scene 2.

Thersites. . . . it will not in circumvention deliver a fly from a spider, without drawing the massy irons, and cutting the web.
TROILUS AND CRESSIDA, Act ii. Scene 3.

Mercutio. Her traces of the smallest spider's web;
ROMEO AND JULIET, Act i. Scene 4.

GNAT.

Ant. S. When the sun shines let foolish gnats make sport,
But creep in crannies when he hides his beams.
COMEDY OF ERRORS, Act ii. Scene 2.

Biron. O me, with what strict patience have I sat,
To see a king transformed to a gnat!
LOVE'S LABOUR'S LOST, Act iv. Scene 3.

Bassanio. Faster than gnats in cobwebs.
MERCHANT OF VENICE, Act iii. Scene 2.

Hubert.   Come, boy, prepare yourself.
Arthur. Is there no remedy?
Hubert.   None, but to lose your eyes.

Arthur. O heaven!—that there were but a mote
 in yours,
A grain, a dust, a gnat, a wandering hair,
Any annoyance in that precious sense!
KING JOHN, Act iv. Scene 1.

Clifford. And whither fly the gnats but to the sun?
KING HENRY VI., Part III. Act ii. Scene 6.

Cleopatra. . . . the flies and gnats of Nile
ANTONY AND CLEOPATRA, Act iii. Scene 11.

Mercutio. Her waggoner a small gray-coated gnat,
ROMEO AND JULIET, Act i. Scene 4.

Simonides. . . . like to gnats,
Which make a sound, but kill'd are wonder'd at.
PERICLES, Act ii. Scene 3.

Imogen. I would have broke mine eye-strings;
 crack'd them, but
To look upon him; . . . . .
. . . .
Nay, follow'd him, till he had melted from
The smallness of a gnat to air;
CYMBELINE, Act i. Scene 3.

CRICKET.

Mamilius. I will tell it softly;
Yon crickets shall not hear it.
A WINTER'S TALE, Act ii. Scene 1.

Petrucio. . . . thou winter-cricket thou:
TAMING OF THE SHREW, Act iv. Scene 3.

Prince Henry. Shall we be merry?
Poins. As merry as crickets, my lad.
>> KING HENRY IV., Part I. Act ii. Scene 4.

Mercutio. Her whip of cricket's bone;
>> ROMEO AND JULIET, Act i. Scene iv.

Gower. And crickets sing at the oven's mouth,
>> PERICLES, Act iii. GOWER.

Jachimo. The crickets sing, and man's o'erlabour'd sense
Repairs itself by rest.
>> CYMBELINE, Act ii. Scene 2.

Lady Macbeth. I heard the owl scream, and the crickets cry.
>> MACBETH, Act ii. Scene 2.

BEETLE.

Isabel. . . . the poor beetle that we tread upon,
In corporal sufferance finds a pang as great
As when a giant dies.
>> MEASURE FOR MEASURE, Act iii. Scene 1.

1st Fairy. Beetles black, approach not near;
>> MIDSUMMER NIGHT'S DREAM, Act ii. Scene 2.

Edgar. The crows, and choughs, that wing the midway air,
Show scarce so gross as beetles;
>> KING LEAR, Act iv. Scene 6.

Belarius. And often, to our comfort, shall we find
The sharded beetle in a safer hold
Than is the full-wing'd eagle.
CYMBELINE, Act iii. Scene 3.

Macbeth. . . . ere, to black Hecate's summons,
The shard-borne beetle, with his drowsy hums,
Hath rung night's yawning peal,
There shall be done a deed of dreadful note.
MACBETH, Act iii. Scene 2.

FLY.

Orlando. I would not have my right Rosalind of this mind; for, I protest, her frown might kill me.
Rosalind. By this hand, it will not kill a fly.
AS YOU LIKE IT, Act iv. Scene 1.

Autolycus. . . . in the hottest day prognostication proclaims, shall he be set against a brick wall, the sun looking with a southward eye upon him, where he is to behold him with flies blown to death.
A WINTER'S TALE, Act iv. Scene 3.

Falstaff. . . . for thy walls,—a pretty slight drollery, or the story of the prodigal, or the German hunting in water work, is worth a thousand of these bed-hangings, and these fly-bitten tapestries.
KING HENRY IV., Part II. Act ii. Scene 1.

Clifford. The common people swarm like summer flies:
KING HENRY VI., Act ii. Scene 6.

Iago. And though he in a fertile climate dwell,
Plague him with flies :
                   OTHELLO, Act i. Scene 1.

Cominius. . . . . . .
Or butchers killing flies.
                   CORIOLANUS, Act iv. Scene 6.

Marina. I never kill'd a mouse, nor hurt a fly :
I trod upon a worm against my will,
But I wept for it.
                   PERICLES, Act iv. Scene 1.

Flavius. . . . one cloud of winter showers,
These flies are couch'd.
                   TIMON OF ATHENS, Act ii. Scene 2.

Arviragus. . . . smiling, as some fly had tickled slumber,
                   CYMBELINE, Act iv. Scene 2.

Titus. What dost thou strike at, Marcus, with thy knife?
Marcus. At that that I have kill'd, my lord ; a fly.
Titus. Out on thee, murderer ! thou kill'st my heart ;
   . . . .
   . . . .
Marcus. Alas, my lord, I have but kill'd a fly.
Titus. But how, if that fly had a father and
     mother?
How would he hang his slender gilded wings,
And buzz lamenting doings in the air !
Poor harmless fly !
That, with his pretty buzzing melody,

Came here to make us merry; and thou hast kill'd him.
TITUS ANDRONICUS, Act iii. Scene 2.

Lady Macduff. Sirrah, your father's dead;
And what will you do now? How will you live?
Son. As birds do, mother.
Lady Macduff. What, with worms and flies?
MACBETH, Act iv. Scene 2.

FLEA.

Mrs. Ford. If you find a man there, he shall die a flea's death.
MERRY WIVES OF WINDSOR, Act iv. Scene 2.

Petrucio. Thou flea,
TAMING OF THE SHREW, Act iv. Scene 2.

2d Carrier. I think this is the most villanous house in all London road for fleas:
KING HENRY IV., Part I. Act ii. Scene 1.

Boy. Do you not remember, 'a saw a flea stick upon Bardolph's nose;

Orleans. —— that's a valiant flea, that dare eat his breakfast on the lip of a lion.
KING HENRY V., Act ii. Scene 3;
Act iii. Scene 7.

CATERPILLAR. CANKER.

Proteus. Yet writers say, as in the sweetest bud
The eating canker dwells, so eating love
Inhabits in the finest wits of all.
    Valentine. And writers say, as the most forward bud
Is eaten by the canker ere it blow,
Even so by love the young and tender wit
Is turn'd to folly;
        Two Gentlemen of Verona, Act i. Scene 1.

    Titania. Some, to kill cankers in the musk-rose buds;
        Midsummer Night's Dream, Act ii. Scene 2.

    1st Servant. . . . the whole land,
Is full of weeds; . . . .
. . . .
. . . her wholesome herbs
Swarming with caterpillars?
        King Richard II., Act. iii. Scene 4.

    Poins. O, that this good blossom could be kept from cankers!
        King Henry IV., Part II. Act ii. Scene 2.

    York. Thus are my blossoms blasted in the bud,
And caterpillars eat my leaves away:
        King Henry VI., Part II. Act iii. Scene 1.

    Lysimachus. . . . a courtesy,
Which if we should deny, the most just gods

For every graff would send a caterpillar,
And so inflict our province.—
PERICLES, Act v. Scene 1.

GRUB.

Mercutio. Her chariot is an empty hazel-nut,
Made by the joiner squirrel, or old grub,
Time out o' mind the fairies' coach-makers.

Friar.   Tell me, good my friend,
What torch is yond', that vainly lends his light
To grubs and eyeless skulls?
ROMEO AND JULIET, Act i. Scene 4;
Act v. Scene 3.

INDEX.

Aconitum, 68.
Acorn, 52.
Adder, 193.
Alligator, 186.
Anchovy, 179.
Ant, 206.
Ape, 171.
Apples, 44.
Apricot, 41.
Asp, 199.
Aspen, 31.
Ass, 127.

Baboon, 173.
Balm, 62.
Barley, 72.
Barnacle, 182.
Bat, 188.
Bay, 33.
Beans, 74.
Bear, 165.
Bee, 202.
Beetle, 211.
Berries, 52.
Bilberry, 49.
Blackberry, 48.
Blindworm, 197.
Boar, 139.
Box, 31.

Brambles, 39.
Briers, 39.
Broom, 39.
Buck, 142.
Bull, 130.
Bullock, 130.
Bur, 22.
Burnet, 62.
Butterfly, 206.

Cabbage, 54.
Calf, 132.
Camel, 156.
Cameleon, 187.
Camomile, 68.
Canker, 215.
Carnation, 7.
Carp, 179.
Cat, 150.
Caterpillar, 215.
Cedar, 26.
Cherry, 46.
Chestnut, 50.
Chicken, 111.
Chough, 84.
Clove, 63.
Clover, 23.
Cock, 109.
Cockle, 19.

Cockle, 182.
Cod, 175.
Coloquintida, 68.
Columbine, 8.
Coney, 147.
Conger, 177.
Corn, 69.
Cow, 131.
Cowslip, 16.
Coystril, 83.
Crab, 48.
Crab, 181.
Crab tree, 36.
Cricket, 210.
Crocodile, 185.
Crow, 84.
Cuckoo, 98.
Cuckoo-flowers, 23.
Currants, 51.
Cygnet, 103.
Cypress, 29.

Dace, 178.
Daffodil, 13.
Daisy, 14.
Damson, 43.
Darnel, 19.
Dates, 51.
Daw, 88.

Deer, 140.
Dewberry, 49.
Dive-dapper, 106.
Dock, 20.
Doe, 142.
Dog, 120.
Dog-fish, 184.
Dolphin, 184.
Dove, 92.
Drone, 204.
Duck, 108.

EAGLE, 77.
Eel, 178.
Eglantine, 15.
Elder, 35.
Elephant, 155.
Elm, 28.
Eyas, 82.
Eyas-musket, 82.

FALCON, 79
Fawn, 142.
Fennel, 61.
Fern, 17.
Ferret, 149.
Fig, 43.
Filberd, 105.
Finch, 99.
Fitchew, 174.
Flax, 74.
Flea, 214.
Fly, 212.
Fowl, 112.
Fox, 143.
Frog, 190.
Fumitory, 20.
Furze, 39.

GARLIC, 57.
Gillivor, 9.
Ginger, 63.

Glow-worm, 207.
Gnat, 209.
Goat, 137.
Goose, 107.
Gooseberry, 47.
Gosling, 108.
Goss, 39.
Grapes, 41.
Grasshopper, 207.
Grub, 216.
Gudgeon, 179.
Guinea-hen, 173.
Gull, 105.
Gum, 67.
Gurnet, 177.

HAGGARD, 81.
Handsaw, 104.
Hare, 145.
Harebell, 17.
Harlock, 20.
Hart, 143.
Hawk, 80.
Hawthorn, 38.
Hazel, 34.
Hazel-nut, 50.
Heath, 39.
Hebenon, 67.
Hedgehog, 150.
Hedge sparrow, 100.
Heifer, 132.
Hemlock, 20.
Hemp, 74.
Hen, 111.
Herring, 176.
Hind, 140.
Hips, 39.
Hog, 138.
Holy thistle, 9.
Holly, 36.
Honeysuckle, 15.

Horse, 114.
Humble Bee, 203.
Hyssop, 62.
Hyena, 164.

IVY, 37.

JAY, 89.

KITE, 88.

LADYSMOCK, 15.
Lamb, 135.
Lapwing, 99.
Lark, 95.
Laurel, 35.
Lavender, 9.
Leek, 57.
Lemon, 46.
Leopard, 162.
Lettuce, 56.
Libbard, 162.
Lily, 6.
Ling, 39.
Lion, 157.
Lioness, 160.
Lizard, 187.
Locusts, 68.
Love-in-Idleness, 17.

MACE, 64.
Mackerel, 176.
Magpie, 87.
Mallard, 106.
Mallet, 106.
Mandragora, 67.
Marigold, 8.
Marjoram, 59.
Marmozet, 173.
Martlet, 100.
Medlar, 47.
Minnow, 184.
Mint, 59.

INDEX.

Mistletoe, 36.
Mole, 148.
Monkey, 170.
Moth, 207.
Mouse, 154.
Mulberry, 47.
Mule, 129.
Mushroom, 56.
Muskcat, 174.
Mussel, 182.
Mustard, 65.
Myrtle, 34.

NETTLE, 18.
Newt, 197.
Nightingale, 93.
Nutmeg, 63.
Nuts, 49.

OAK, 24.
Oats, 72.
Olive, 33.
Onion, 56.
Orange, 45.
Osiers, 38.
Osprey, 104.
Ostrich, 75.
Otter, 174.
Ousel-cock, 97.
Owl, 83.
Ox, 131.
Oxlip, 14.
Oyster, 180.

PADDOCK, 190.
Palm, 32.
Pansy, 8.
Panther, 163.
Pard, 162.
Parrot, 76.
Parsley, 55.

Partridge, 90.
Peacock, 75.
Pears, 43.
Peas, 73.
Peascod, 55.
Pelican, 105.
Pepper, 65.
Peppercorn, 64.
Pheasant, 90.
Pig, 138.
Pigeon, 92.
Pike, 178.
Pilchard, 177.
Pine, 30.
Pink, 9.
Plantain, 23.
Plum, 42.
Polecat, 174.
Pomegranate, 36.
Poor John, 177.
Poppy, 67.
Porcupine, 170.
Porpus, 183.
Potato, 54.
Prawns, 181.
Primrose, 13.
Prunes, 51.
Puttock, 83.

QUAIL, 90.
Quince, 45.

RABBIT, 147.
Radish, 54.
Raisins, 52.
Ram, 133.
Rat, 152.
Raven, 86.
Reremouse, 188.
Rhinoceros, 164.
Rhubarb, 66.
Rice, 74.

Robin, 96.
Roe, 143.
Rook, 87.
Rose, 1.
Rue, 60.
Rush, 21.
Rosemary, 61.
Rye, 73.

SAFFRON, 64.
Salmon, 175.
Samphire, 56.
Savory, 59.
Scamel, 105.
Scorpion, 196.
Senna, 66.
Serpent, 190.
Shark, 183.
Sheep, 133.
Shrimp, 181.
Slug, 201.
Snail, 200.
Snake, 194.
Sow, 139.
Sparrow, 100.
Spider, 208.
Squirrel, 148.
Stag, 139.
Stannyel, 82.
Starling, 89.
Strawberry, 40.
Swallow, 99.
Swan, 103.
Sycamore, 28.

TENCH, 179.
Thistle, 20.
Throstle, 97.
Thrush, 97.
Thyme, 62.
Tiger, 161.
Toad, 188.

Tortoise, 186.
Trout, 175.
Turkey, 109.
Turnip, 54.

UNICORN, 164.

VETCHES, 73.
Vine, 37.
Violet, 11.

Viper, 196.
Vulture, 79.

WALLFLOWER, 9.
Walnut, 50.
Wasp, 205.
Weasel, 149.
Whale, 182.
Wheat, 71.
Wild duck, 106.

Wild fowl, 106.
Wild goose, 107.
Willow, 31.
Wolf, 167.
Woodbine, 15.
Woodcock, 91.
Worm, 197.
Wren, 102.

YEW, 28.

THE END.

Printed by EDWIN SLATER, *Manchester.*

www.ingramcontent.com/pod-product-compliance
Lightning Source LLC
Chambersburg PA
CBHW022013220426
43663CB00007B/1059